STALLION MANAGEMENT

A Practical Guide for Stud Owners

Also by A. C. Leighton Hardman

THE AMATEUR HORSE BREEDER

STALLION MANAGEMENT

A Practical Guide for Stud Owners

Written and illustrated by

A. C. Leighton Hardman

PELHAM BOOKS

First published in Great Britain by Pelham Books Ltd
52 Bedford Square, London WC1B 3EF
1974

ISBN 0 7207 0699 8

Set and printed in Great Britain by
Tonbridge Printers Ltd, Peach Hall Works, Tonbridge, Kent
in Baskerville eleven on thirteen point on paper supplied by
P. F. Bingham Ltd, and bound by James Burn
at Esher, Surrey

To everyone both human and equine
who helped me write this book

CONTENTS

ILLUSTRATIONS

Note: Some of the photographs of foaling appeared previously in THE AMATEUR HORSE BREEDER.

DRAWINGS

ACKNOWLEDGEMENTS

I should like to express my gratitude to Dr L. B. Jeffcott and Sidney Ricketts, M.R.C.V.S. for their kindness and helpfulness in reading and correcting the manuscript of this book, also to Christopher May, M.R.C.V.S. for his helpful advice. I would also like to thank Bill Cowell and everyone at Cheveley Park Stud for all they taught me and especially for allowing me to photograph the stud, also to Bill Johnson, Derisley Wood Stud, for allowing me to photograph his stallions at work, and to Patricia Lindsay for allowing me to photograph her beautiful Arab stallion Gerwazy. Peter Bolton for his advice on Agricultural matters, Mummie for reading and improving the manuscript and Peter Faulkes, for his helpful advice on photographic matters.

NEWMARKET

MARCH 1974

A.C.L.H.

FOREWORD

By Clive Graham

The reader will not be required to probe deeply into this book before realising that it is the work of a writer who has acquired her expertise in the hard school of practical experience.

Ann Leighton-Hardman has crammed this knowledge inside less than a score of years since early days at her family home near Preston in Lancashire, where she earned precocious fame as a wheeler-dealer in children's ponies. In her late teens, she ventured £300 from her savings into the purchase of a colt from the Briery Close Arabian Stud, near Windermere. This colt earned her £450 each year as a stallion before she re-sold him for £800, after deciding to leave the family home for a job in Newmarket.

For a season she worked as secretary at the Aislabie Stud, run by the former jockey Dick Perryman. She moved then to the Equine Research Station at Balaton Lodge, Newmarket, completing an eighteen-months' stint and finishing it as lay assistant to W. R. ('Bob') Cook, F.R.C.V.S., ear, nose and throat specialist.

Hooked now to the thoroughbred scene of operations, Ann switched her employment to work for Richard Galpin's Newmarket Bloodstock Agency at their office near Tattersalls' Sale Paddocks. Her principal assignment here was to trace and make complete statistics of pedigree-charts for their international clientele.

At the time of writing this book, Ann had completed another grasshopper feat, gaining the post of assistant manager to Richard Stafford-Smith's Cheveley Park Stud, where Isinglass once reigned supreme nearly a century ago, and where Forlorn River has latterly carved such a name for himself as a sire of winners.

Ann's education – let me not forget – included two years at the Lancashire County Agriculture College. The conclusions and recommendations contained in this book come, therefore, from a well-trained head on young and sturdy shoulders.

To the seasoned professional horse breeders, there may be nothing of eye-catching impact, but the writer dots all the 'i's' and crosses all the 't's'. For this reason alone, it deserves its place on the shelves of every stud-farm office, as a fount of reference. For the amateur 'do-it-yourself' breeder, this book should prove an almost indispensable manual. In spelling out all the manifold problems associated with horse breeding and their sensible solutions, Ann Leighton Hardman has performed a task which may not achieve any great financial reward, but will still be treasured and appreciated by those who recognise its sound, solid, basic worth in this esoteric field.

Clive Graham.

I

INTRODUCTION TO STALLION OWNERSHIP

Having a stallion of your own at stud is one of the most exciting and absorbing aspects of horse breeding and so much more interesting than just owning mares. Basically there are two types of stud which stand stallions:

(i) The private stud: one which keeps a stallion just for the owner's use.

(ii) The public stud: one which keeps a stallion or stallions at stud for use by mare owners in general.

Running a private stud which stands a stallion does mean that you will have the joy of foaling all your mares at home and of watching the baby foals grow up which is often missed when the mares have to go away to stud. There is the added advantage of a greatly reduced disease risk as the mares and young foals never need to leave the stud during this critical period in the foals' lives. With the rising cost of keep at public studs there is a great saving in stud bills *but* the stallion must be chosen carefully and *must* be of equivalent merit to those normally used at a public stud, otherwise the standard and ultimate marketability of the young stock will suffer.

A public stud can be the greatest fun; it will enable you to meet lots of interesting people and make many new friends as well as having the pleasure of handling different types of mares and their foals. There is an enormous amount of satisfaction to be gained from getting numbers of mares in foal after dealing with their various eccentricities and nothing can quite equal the joy of seeing your stallion's foals the following spring.

An average stallion should be able to pay for himself in two seasons without any difficulty, which will leave you with your own nominations at the cost of keep, depreciation, etc., which when spread over several mares will amount to very little as compared with normal stud fees.

It is usual for an adult stallion to have about forty mares booked to him each season, and of these nearly half could have foals at foot and might require to be brought in at night. Some few small ponies which may be prone to laminitis may also have to be kept in or left on very bare pasture.

In each case you will need at least twenty loose-boxes per stallion or you could possibly manage with slightly less if you have some covered yard space which can be used for some of the barren or maiden mares. You will need at least forty acres per stallion to provide enough grass during the short period in the year when all the visiting mares are on the stud. After this many studs run a herd of beef cattle to keep the grass down during the remaining months.

A big responsibility rests with every stallion owner, as careful choice of the individual horse is essential. A mare normally has only one foal each year whereas a stallion can sire upwards of thirty foals. Therefore a bad horse can do untold damage to his breed and to the horse population of the area in which he stands. Because of this, the Ministry of Agriculture Fisheries and Food demand that every colt over the age of two years shall have a licence or permit, with certain exceptions which are:

(i) Thoroughbreds – ie. horses entered or eligible for entry in the *General Stud Book*.

(ii) Certain breeds of pony turned out or standing in an

area where ponies of the breed usually run on free range, providing the stallion is not travelled or exhibited for service.

(iii) Shetlands.

A stallion licence is required as soon as the colt is two years old. For this purpose, his age is calculated from January 1st of the year of birth. Application for a licence should be made to:

Livestock Improvement Branch,
Ministry of Agriculture, Fisheries and Food,
Great Westminster House,
Horseferry Road, London SW1.
Or in Scotland to:
Broomhouse Drive,
Edinburgh 11.

Application should be made between 1st July and 30th September of the colt's yearling year. The current licensing fee is £11.55 which is not returnable if the licence is refused.

Before the licence is issued the Ministry will send a veterinary surgeon appointed by them, to look at the horse. He will examine the horse first in his loose box and then outside, when he will need to see him both trotted up in hand and lunged at the canter, or in the case of broken horses, ridden at fast work. So it is essential to have the colt quiet to handle and to have taught him to lunge before the licence is applied for.

The veterinary surgeon will want to check the following points:

(i) Any sign of contagious or infectious disease.

(ii) Evidence of any possible hereditary conditions and weaknesses such as:

Cataract	Ringbone (high and low)
Whistling or roaring	Bone Spavin
Side-bone	Shivering
Navicular disease	Defective genital organs
Stringhalt	

A stallion over nine years old will not on examination

have his licence revoked if it has been in force for at least two years. Any horse may have his licence revoked for inadequate fertility if this can be proved. If for any reason the stallion fails on inspection, the owner may appeal on payment of £16.80, which will be returned in the event of the appeal being turned down. In the case of an appeal the stallion will be re-examined by a referee; if he is again turned down the owner will be instructed to have him castrated or slaughtered.

A permit may be issued when the Ministry decides that for some reason a licence should be temporarily refused but that the stallion should be kept entire. It is usually made a condition that the stallion should not be used at stud until a full licence is granted.

It is a legal requisite that the stallion owner must produce the licence if he is asked to do so by a police officer, Ministry official, or person in charge of a visiting mare.

When the stallion is sold or leased for a period longer than six months, the licence should be sent to the Ministry for transfer to the new owner. Should the stallion remain in the same ownership but he moved to a new address, the Ministry must be notified.

Although a colt is licenced at two years old this does not mean that he should be used for stud at this early age. Like a two-year-old filly, a colt is still growing and developing, so to use him for stud duties at this time could stunt his growth and damage his constitution for the rest of his life. He can, however, be used for a few mares as a three-year-old and commence full stud duties as a four-year-old, by which time he will virtually have finished growing.

When you first decide to keep a home-bred colt or buy a prospective stallion to stand at public stud, it is essential that you should find out if there will be a demand in your particular area for the type or breed of horse you have in mind. If there are numerous stallions of a breed already at stud in your area, you may find that the local demand is already saturated and it may be wiser to stand a horse of another breed or type. Unless, of course, you are specifically buying in a breed or blood line you wish to use on your own mares.

For instance, if there are numerous Arabian stallions, buy a small Thoroughbred horse instead. But above all make certain that you do not buy a stallion which is very closely related to a horse already at stud in your area, otherwise you may find that breeders will be unable to send you any mares by that horse, which are consequently also related to your stallion. This factor would reduce your potential market considerably.

This is also a point when a stallion is used on a private stud and thought must be given to the time when the majority of the young fillies on the stud are all by one particular stallion.

When buying your stallion, some consideration should be given to the stud fee which you can reasonably charge. It is often considered the best plan to stand a horse at a low fee for the first two seasons and choose the best mares from those which are offered. If on the other hand you have several good mares of your own to send to the horse, it would probably be advisable to stand him at a higher fee, as this would tend to raise the market value of your foals and you would need only a few outside mares to complete his book for the season. At all costs, do not be tempted to over-price the stallion, otherwise you will never get any mares to him and one day you may have to lower his fee, which is never good for publicity!

See how much similarly bred horses which stand in your area cost, and then price your horse and the keep fees accordingly. It is always better to have to put a fee up due to excessive popular demand for the horse, than ever to have to bring it down because you cannot sell the nominations.

Stud fees are usually divided into four categories:

(i) Straight fee: no concessions are given, the fee is payable regardless of whether the mare is in foal.

(ii) No foal, free return: this is self-explanatory but the free return must be taken the following year, otherwise you might find that you have about thirty free-return mares in one season!

(iii) No foal, no fee: in this case, it must be stipulated that the mare is tested barren by October 1st in the year of covering, otherwise the fee is due. A certificate of barrenness must be signed by a veterinary surgeon and dated before October 1st if the fee is not to be charged.

(iv) No live foal, no fee: the stud fee is usually payable at the end of the stud season, unless the mare is tested barren. The fee is returnable in full, if the mare does not produce a live foal which stands and survives for forty-eight hours. Very occasionally a stallion is advertised as payment of the stud fee after a live foal is produced.

It is really a matter of personal preference which of the above categories you choose, but several factors must be taken into consideration. Most mare owners would rather patronize a horse which stands with a *no foal, no fee* concession, since with the high cost of producing a foal these days, this concession and the last represent the least risk. However, the national foaling average for Thoroughbred mares is only two foals for every three years that a mare is at stud, therefore in order to compensate for loss of income from mares which fail to conceive, the stud fee must be raised by one third.

It is probably fairer to stand a stallion with the *no foal, no fee* concession for his first season until his fertility has been proved. The *no live foal, no fee* concession relies to a large extent on factors outside the control of your stud which is sometimes not a sound policy.

Many stallions stand at stud with the option of either a *straight fee* or the *no foal, no fee* concession, the fee being slightly higher in the latter case as explained above. With most high priced Thoroughbred stallions, no concessions are given. It should also be stated on both the stud card and nomination forms that the concessions apply to approved mares only, otherwise you may find yourself having to accept a whole crowd of old or permanently barren mares for your horse, which will give a false impression of his fertility.

Above all, you must really believe in the stallion you choose; if you do not think that he will be a great success at stud you can rest assured that no one else will. You will find that it is largely your own enthusiasm for the horse which will sell the nominations in the first few years he is at stud, at least until he has proved himself one way or the other.

2

STALLION PROMOTION

The stallion owner or stud manager, must, above all, be a good salesman – then everything else should follow.

First impressions always count, so the entrance to your stud should always look neat and tidy. All the paddocks on the stud should be well maintained as far as possible and should have post and rail fencing, or neatly clipped hedges. The grass should be kept topped regularly, so that the general effect is good.

The stallion's loose-box should be immaculate; you can to a certain extent influence a horse's appearance by his immediate surroundings. A large airy, clean loose-box with all the paint work in good condition, will make a horse look a much better individual than a dark, dingy, cobweb infested loose-box. This works in very much the same way as a high class frame does for an oil painting.

Similarly, you should keep the stallion's tack clean and polished at all times and his feet in good order. Ideally, he should be well groomed and shining before any visitor sees him, but there is always the occasional person who arrives, without first telephoning to make an appointment. If the horse is groomed thoroughly every day he should not look too bad if somebody does turn up unexpectedly. This, of course, only applies to the stabled stallion and the horse which is turned out during the day. Stallions running with their mares should, where possible, be caught up for visitors to see as this is preferable to looking at them in the middle of a paddock on what might be a cold wet day.

In the case of the stabled horse the type of system you choose will, to a large extent, have to depend on the individual horse. Ideally stallion boxes should be located in the quietest part of the stud, but in such a position so that the horse has something to look at while he is stabled. This should help prevent him from getting bored and developing a stable vice, in order to entertain himself!

An ideal layout from this point of view is shown in plate 12. Here the stallions are housed in a quiet yard, looking out on to a belt of trees, which is preferable to the conventional stallion yard with a high wall all round.

Another system which usually works well, particularly when there is only one stallion on the stud, is to build a loose-box at one end of a 3-acre paddock. Under this system, the loose-box door can be left open all day and the horse able to wander in and out at will. The gateway into the paddock must be made large enough to allow a tractor and trailer through so that standard paddock maintenance of topping, etc., can take place (see Fig. 1).

It is convenient to have an extra door in the back of the loose-box, so that visitors will not have to go though the paddock in order to see the horse. The only disadvantage to this system of management is that it does not make any provision for a change of pasture for the horse.

Fencing round stallion paddocks can be the conventional close boarding (see plate 8) or ordinary post and rails which can be seen in plate 6. In this case, the fencing should be made a little higher and stronger than normal in the interests of safety. Most stallions, with time, will settle down well in a post and rail paddock all the year round, even during the covering season. Nothing improves the general impression of a stud more than white paddock fencing, and wherever

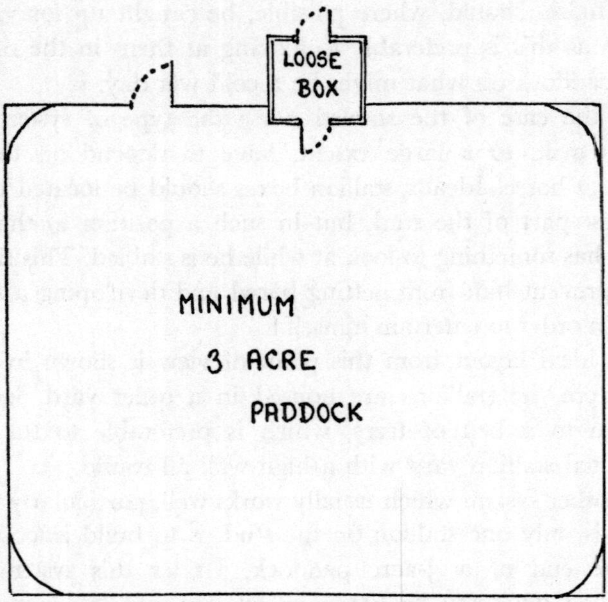

Fig. 1. Paddock and loose box

possible this should be done, particularly round the stallion paddocks.

It is essential that every stallion should have some form of stud card (a brochure giving details of his breeding and performances). This can then be sent out to all prospective visiting mare owners, when they enquire about the horse. Therefore, it is essential that the card is so designed that it will sell the nominations. A stud card should be produced the moment a stallion retires to stud, so that it is ready by the time the first advertisement for the horse appears.

When designing stud cards the best thing is to get the stud cards of as many stallions as you can – not necessarily just the breed or type you are interested in – study them all well and then try to design one which is different. The reason for this is that when people are looking for a stallion to which they can send their mare, they will write round to several stallion owners for stud cards. If yours is different from everyone else's it will make a bigger impact. Don't be tempted to design a cheap stud card as this will only have the effect of undervaluing your horse.

As already stated in Chapter 1, you must believe in your own stallion and this belief should be reflected in the stud card. Do not be frightened to claim that your stallion comes from the greatest sire line in the world or is the best mover in the country, etc., if you have any grounds for believing this to be so. If you believe it enough yourself, you will find other people will begin to agree with you.

Stud cards should always contain a posed photograph of the horse, although it is better to leave out the photograph altogether than to use a bad one. This would do far more harm than good, as a bad photograph will only tend to put off people to the extent that they will probably not even bother to come to look at the horse. If your stallion has only just retired to stud for the first time and has not had time to let down, it would probably be better to use a photograph of the horse's head only.

You should always carry some stud cards and nomination forms with you wherever you go. Put some in your car; you will never know exactly when you might meet someone who is interested in sending a mare to your stallion, and then you can get them signed up on the spot.

It always pays to advertise, particularly when you are trying to promote a new stallion. The more people who know about your horse the better. You should put an advertisement in the national equine press in the autumn and repeat it at least every month until half-way through the stud season or until the horse is full, which ever comes first. You should then announce that he is full – which is always good for publicity. It is also advisable to advertise extensively in your local press – a local farming paper is often a good medium. Keep the advertisement simple; it is usually much more eye-catching if it is instantly easy to read.

Decide at the beginning of the season approximately how

many mares you intend to accept to your horse. In the first season, it is essential that the stallion should have sufficient mares so that he will have enough progeny in his first crop to make his name. If by any chance you cannot get enough outside mares to your horse, it is a very good idea to buy some suitable mares to make up the number. Later, you can enter them for selected horse sales throughout the country. If you have bought the mares wisely, the resulting progeny should help to advertise your horse extensively. The wider the progeny are scattered the greater will your potential market be in future years, as nothing advertises a horse better than his progeny if they are good. This can be taken to the extreme and great efforts can be made to cultivate an export market for your horse. Even covering to southern hemisphere times (in most countries this is August 15th to January 15th for Thoroughbreds) can be contemplated, although you will be stuck with the progeny if your scheme does not work.

It is usually sound policy to accept any really good mares in the horse's first season, on a reduced fee basis – even giving a free nomination to an outstanding mare, since a really top class foal in the stallion's first crop will do more than anything else to make him for several seasons.

If your main aim is to produce stock for the show ring, it is really up to you to make sure that your stock get there. Therefore, be prepared to offer a service to your mare owners, give them advice when they ask you to, but be careful not to push yourself forward and advise people when they do not want it. This will only put people off using your horse in future years. If asked, be prepared to get the progeny ready for the show ring yourself and even show them, for the people concerned. It is sometimes a good idea to think of competitions which will help to promote your stallion: you could offer a prize for the young horse by your stallion which gains the most awards at shows throughout the season, giving the prize on a points system. The prize could either be a trophy, money or a free nomination; the latter will have the added advantage of possibly getting back the dam of the good prize winner, for your stallion, so that she can have another foal by him – all good publicity for the horse!

Where you are dealing with non-pedigree stock, it is a very good idea to have cards printed, so that mare owners can fill in the details of the progeny out of the mares they sent to your stallion. These should be sent stamped so that you will be fairly certain of getting them back. You can put in a clause asking for comments on the progeny which, with the owner's consent can be used for advertising purposes.

There are four stallion guides, published annually in which one can advertise. These are:

For Thoroughbred Stallions only
The Stallion Review,
26 Charing Cross Rd.,
London WC2H ODJ.

Sires for 19.
Stud and Stable Ltd.,
149 Fleet Street,
London EC4A 2BU.

For any horses which are likely to breed hunters, including Thoroughbred stallions
Hunter Stallion Directory,
Vardon Publications,
Meretown Stud,
Newport, Shropshire.

For all pony stallions
Pony Stallion Guide,
Vardon Publications,
Meretown Stud,
Newport, Shropshire.

NOTES ON PHOTOGRAPHING THE STALLION

A bad photograph of a horse is worse than no photograph at all. When advertising stallions at stud it always helps to use a good photograph of the horse, since many people will look through the various stallion guides and reviews to choose a mate for their mare. When doing this, they will probably be looking for a horse which is particularly strong in the points of conformation where their mare is weak. A good clear photograph may help to sell nominations for you, whereas a bad photograph will only tend to put off people and might actually loose nominations which would otherwise have been sold. If you cannot get a really good posed photograph, it is better to use a head study or an action photograph – in the case of a horse out of training, a good racing photograph is ideal – and hope that people will come to see the horse for themselves.

For the photograph, the stallion should be well-groomed, his mane and tail brushed out and all his mane put neatly over on to one side (usually the off-side) of his neck. If the mane is uneven and a few pieces stick up, these should be wetted and brushed down or sprayed with lacquer. He should be wearing a clean, shining stallion bridle; a headcollar does not give the right impression of efficiency to the prospective mare owners. Your advertisement constitutes a shop window for BOTH the horse and stud. For this reason, be very careful when choosing a location for the photograph – make sure it looks neat and tidy!

If the horse is wearing an ordinary bridle with reins, these are better put behind the withers rather than on the neck, as the latter will only tend to make his front look short.

It is advisable to mow a strip of grass in a suitable paddock. If you stand your horse in long grass, it will tend to hide his feet and pasterns. It is often a good idea to photograph a horse against the sky-line, so that his outline is not masked by vegetation or buildings. Alternatively, a plain solid wall makes a good background. If you can, choose a calm day;

when the wind is blowing, it is sometimes difficult to keep down the mane or prevent the tail from either flowing out behind or covering his hocks.

Dark-coloured horses should be taken against a light background – the sky is ideal or a light-coloured wall. Take very light-coloured horses against a dark background, for instance a high hedge or thick trees, but great care must be taken to ensure that you do not have something growing out of your horse's back in the finished picture!

Ideally the horse should stand with his ears pricked, and to achieve this you will require a third person to entertain him. One of the best ways ways to attract his attention is to use a wireless. Most horses will become interested if you suddenly turn on some music.

The camera should be aimed at a point immediately behind the shoulders and half way down the horse's barrel. If the camera is held higher than this, it will give the impression of a big body on short legs and will also tend to make the horse look smaller; similarly, if the camera is held low down, an impression of long legs is achieved. The sun should be directly behind the photographer, which will ensure that no part of the horse is obscured by dark shadow.

The horse should have his weight evenly distributed on all four legs and should stand with all his legs visible in the finished photograph. For this reason, he should be made to stand with the hind leg furthest from the camera slightly forward, but not so far in advance that there is daylight between the hocks. The foreleg nearest to the camera should be in full support with the other front leg slightly back.

3

MANAGEMENT OF THE VISITING MARES

The official Thoroughbred stud season starts on February 15th and ends on July 15th; but, due to the fact that many ponies and hunters are foaled and kept outside all the year round, the non-Thoroughbred stud season extends into late summer.

With the high cost of keep on public studs, many more people are foaling their mares at home and only bringing them to the stud on or about the sixth day after foaling or at the second oestrus. Maiden mares (those which have never been covered before in their lives), are probably better not accepted too early in the year, as they are notoriously slow in starting to function properly and will often only produce a good egg until the warmer weather comes. However, a few maidens will hold to an early service.

Barren mares (those which have had at least one foal before but which are currently not in-foal) should be accepted as soon as possible, to give them every chance of getting in foal, as some barren mares will inevitably be infected at the beginning of the season and will therefore need cleaning up before they can be covered. Then there is the mare which an owner believes to be in-foal, which will not arrive until at least the second half of the covering season, after she has subsequently proved to be barren.

As each mare arrives at the stud she should be checked to make sure she is healthy and not suffering from any contagious disease. If the mare is coming from abroad or from an area where it is known that there is some possibility of infection, she should be isolated immediately on arrival, looked after by one man alone (an old age pensioner is ideal for this job), until your veterinary surgeon is satisfied that there is no risk of infection.

The condition of the mare should be noted on arrival and the owner informed of any injury which may have occurred during travelling. Over-fat mares are notoriously difficult to get in foal, thin mares in improving condition are usually easy to get in foal providing they are clean and not thin due to worms or some illness. Once they are in foal they will often tend to put on flesh rapidly, and this is particularly noticeable in the case of maiden fillies – especially those straight out of training.

Any mare which arrives at the stud without her name on the headcollar should have it put on, ideally by means of an engraved brass plate, so that the staff can identify the mare easily and get to know her. This is most important as much time can be wasted during the season if some of the mares have no names on their headcollars when a stud-worker is sent to a paddock to catch a visiting mare which he does not know. The cost of the plate can be charged to the owner's account.

Some studs have leather tabs made by their saddler on which they stamp the stallion's name, but with the rising cost of materials this practice is becoming rather expensive. A far cheaper method of recording which stallion a mare is visiting (when there is more than one stallion on the stud), is to work out a colour code for each horse (stallion) and buy a quantity of plastic hen rings to correspond with the colours chosen. A hen ring is then fixed to the near side of the mare's headcollar, and gives an 'at a glance' guide, even when she is some distance from you in the paddock.

In the case of mares which are stabled at night, ideally each mare should be kept to one loose-box during the period of her stay at the stud, so that a routine can be established for both staff and horses. In order to identify the loose-box and act as a guide for the staff and visitors, it is customary to attach a card to the door or adjoining wall, giving all the details of the mare. See Fig. 2.

OWNER ....

MARE ....

SIRE..

DAM ....

IN FOAL TO .... DUE ....

PRODUCE .... FOALED...........................

GOING TO ....

Fig. 2. A stable-door card

These cards can be obtained quite cheaply from Tindall & Son stationers, Newmarket, and will slip into plastic envelopes, which help to protect the card and keep it clean; standard metal card holders are also available. The information should be printed neatly in waterproof Indian ink or punched out on Dymo tape. Typewritten words are often difficult to read, especially towards the end of the season when the type is beginning to fade; red type, in fact, tends to fade out altogether when exposed to the sun.

The name of the stallion the mare is visiting can be done on Dymo tape of the same colour as the hen rings; this helps the staff to identify the loose-boxes, when bringing the mares in from the paddocks.

Basically you will have four groups of visiting mares on the stud, but where the mares range from small ponies to large horses these should be sub-divided, the big mares being put in a different paddock from the small ones to prevent any chance of bullying. The four groups are: in-foal; with foal at foot; barren; and maiden – young fillies which have never been covered before. Older maiden mares can be put with the barren mares if necessary.

At the time of booking the nomination, the mare owner should be asked to complete a nomination form and questionnaire relating to the mare he intends to send. The questionnaire enables the stallion owner to select the best mares from those offered should he so wish (Fig. 3).

As far as possible, people should be discouraged from sending their mares on Sundays, so that the staff, who on a small stud are probably working a seven-day week at this time of the year, can have a little time off.

If your stallion tends to book up quickly each year, a clause can be put into page 1 of the nomination form to the effect that the form should be returned within a fortnight otherwise the booking will not be held. This is important as some people tend to keep the form for a long time or even bring it when they arrived un-heralded with the mare one day.

In cases where the stud fee covers a 'no foal no fee' concession, it is wise to add another clause to the effect that the mare must be available for service up to end of the stud season – i.e. July 15.

As far as possible it is desirable to get the passport or registration number of the mare at the time the nomination is signed as these are often very difficult to obtain from owners at a later date. Where registered mares are concerned, the number must go on the covering certificate.

As soon as the mares arrive at the stud they *must be wormed* and a faecal worm egg count done, (see chapter 16) *before* they are turned out to grass – this is absolutely essential otherwise your paddocks will become 'horse sick' in a very short space of time.

According to the period in the year when they arrive, the

barren mares may be turned straight out to grass, or stabled at night if it is still too cold. Maiden mares should be stabled at night until the warmer weather comes, which will give them a better chance to get in foal early in the year; then

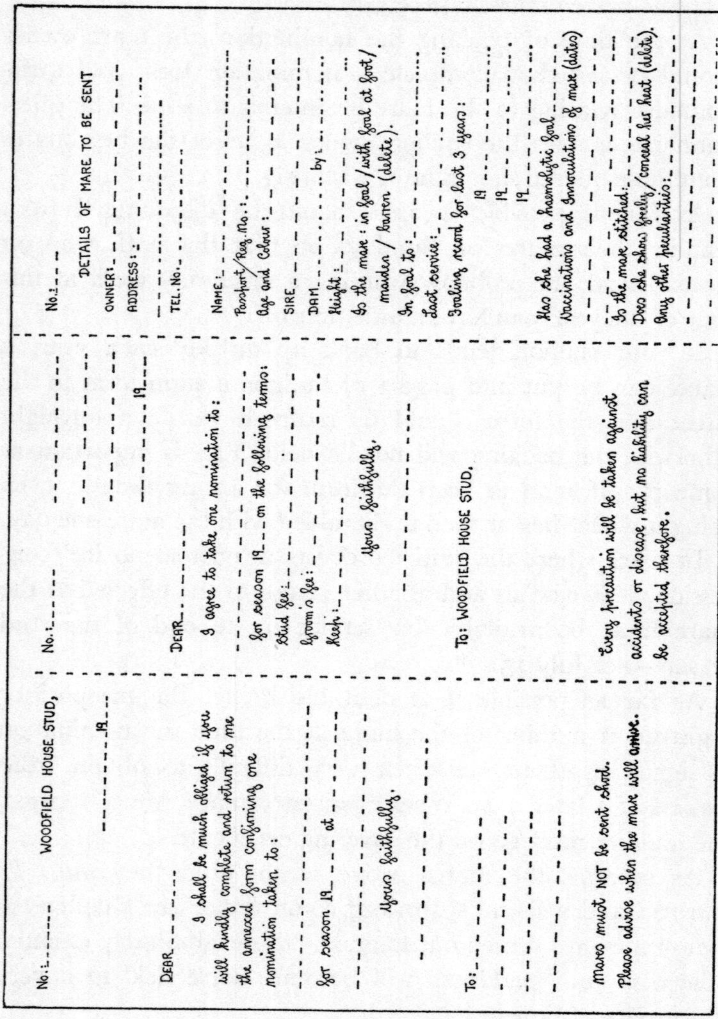

No.:

WOODFIELD HOUSE STUD,

19

DEAR

I shall be much obliged if you will kindly complete and return to me the annexed form confirming one nomination taken to:

for season 19 at

Yours faithfully,

To:

Mares must NOT be sent shod.

Please advise when the mare will arrive.

No.:

19

DEAR

I agree to take one nomination to:

for season 19 on the following terms:

Stud fee:

Groom's fee:

Keep:

Yours faithfully,

To: WOODFIELD HOUSE STUD.

Every precaution will be taken against accidents or disease, but no liability can be accepted therefore.

No.:

DETAILS OF MARE TO BE SENT

OWNER:

ADDRESS:

TEL. No.

NAME:

Passport/Reg. No.:

Age and Colour:

SIRE

DAM by:

Height:

Is the mare: in foal/with foal at foot/maiden/barren (delete).

In foal to:

Last service:

Foaling record for last 3 years:

19

19

19

Has she had a haemolytic foal

Vaccinations and Innoculations of mare (dates)

Is the mare stitched:

Does she show freely/conceal her heat (delete).

Any other peculiarities:

Fig. 3. A sample nomination form

they too can be turned out. Foaling mares should be brought in every night for at least a month before they are due, regardless of breed or type, as ideally, all mares should foal inside in case one has a difficult foaling. Mares with a foal at foot should be kept in at night until the weather is warmer, then at the discretion of the stud owner and according to the breed or type they may be left out day and night.

All the visiting mares on the stud should have their feet trimmed once every four to six weeks. When the ground is hard during the summer it may be found necessary to put front shoes on some mares with soft or brittle feet. Every animal on the stud should be wormed regularly; all the horses running out day and night in the paddocks should be brought up at least once every eight weeks and wormed. This is best done in rotation – paddock by paddock. The mares can then be generally tidied up and any knots taken out of their manes. Ideally all mares which are inside at nights should be brushed over and their manes and tails done each morning before they are turned out. All blacksmith and worming charges should be put on the mare owner's account.

Although you may state on your nomination form that 'no responsibility is accepted for accidents or disease', every stud should make sure that their paddocks, loose-boxes, etc., are absolutely safe for horses. This means that the loose-boxes must be large enough so that animals will not get cast when they lie down and the half-doors high enough so that they cannot jump out. Failing this, all the doors should have metal cages, and the windows should have wire mesh guards to prevent the horses from breaking the glass and cutting themselves badly. For foaling Thoroughbreds and Hunters, it is generally recommended that a loose-box should measure at the very least 16 ft. x 14 ft., but for everyday use 12 ft. x 12 ft. is quite sufficient, and 10 ft. x 12 ft. would be big enough for ponies (see Fig. 26).

It is really only safe to put horses in post and railed

paddocks and young foals should never be put in fields where there is any barbed wire, as they tend to gallop into fences on occasions and could cut themselves badly. Fencing for mares with foals at foot should come low to the ground to prevent the foals from rolling underneath when they lie down near to the fence side (see plate 12), otherwise two rails are sufficient. If at all possible, it is always better to run a double fence between paddocks in order to help to prevent the spread of disease, and so that the horses cannot reach one another over the fence and break it down by leaning on it.

As far as possible, your paddocks should have some form of shelter – either high hedges or trees – as cold winds in the spring will only delay the time when the mares will start to come in-season and ovulate normally. Some people use field shelters, but these must be quite large, otherwise some mares will get kicked, especially when groups of strange mares are turned out together and they all crowd into a shelter away from a cold March wind.

It is essential on any stud that a rotational system of grazing is worked out. Horses are notoriously poor grazers and therefore should be moved from one paddock to the next at least once a month, preferably more often. The paddocks can then be topped in the areas where the horses have not grazed and the whole paddock rested and allowed to grow a little before it is grazed again. The grass must never be allowed to get very long before horses are turned into it, otherwise they will not eat it but only trample it under foot. Ideally cattle should be grazed with horses (see plate 13), which to a large extent, will eliminate any need for topping the paddocks.

However, the following factors should be taken into consideration: all the cattle should be naturally polled or dehorned, to eliminate any risk of injury to the horses and they should be free from Brucellosis, as this disease is transmissible to horses and takes the form of Poll evil, fistulous withers and some forms of soft swellings which can cause obscure lamenesses.

In order to save labour, all water troughs in the loose-boxes and paddocks should be fitted with automatic filling devices but they will need checking daily, to make sure that they are working properly and the troughs are clean. During frosty weather the ice should be broken at least twice a day. It will be found that a long bristle dairy washing-up brush is very useful for scrubbing out water troughs, which should be fitted with plugs so that they can be washed out easily.

4

FOALING

The mare's udder gradually increases in size during the last month of pregnancy until the final week, when it will become shiny and remain large even after exercise. Candles of wax may appear on the ends of each teat (see plate 1). In many cases, the candles of wax disappear and the mares will run their milk from both teats before foaling (see plate 2).

Once the mare's udder becames large, she should be moved into a foaling box, as not all mares will wax up or run their milk before foaling and one should not take any chances particularly with visiting mares.

When you are running a public stud, it is inevitable that some mares will come to be foaled down at your stud, therefore it is essential that you make arrangements for this. You will need at least two foaling boxes, so that one can be rested should a case of disease occur (see Fig. 4). For this reason it is advisable for foaling boxes to be cement-rendered with concrete floors, so that they can be easily disinfected.

However small your stud is, you will need to organise your staff so that someone is sitting-up every night when a mare is near to foaling. On most studs, the men take it in turns to sit up with the foaling mares. For this, they are paid a set amount, which is £8.00 per night at the moment on Newmarket studs. After sitting-up, it is customary to give workers at least the following afternoon off.

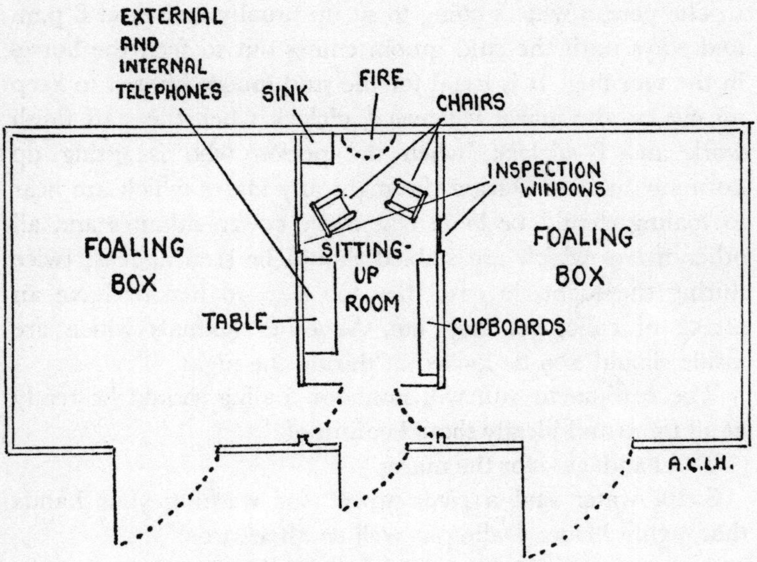

Fig. 4. Foaling unit

The generally accepted system is for students and learners to sit up with experienced staff, until such time as they have seen approximately three foalings when, if they are sufficiently intelligent and mature, they can be allowed to sit up on their own. In every case the person who is sitting-up must call the stud groom, or person running the stud, the moment a mare looks as though she is going to foal – usually when she starts to get warm and certainly by the time she 'breaks water'. Therefore, there should be an internal telephone system between the sitting-up room and the stud groom's house to save time. All foalings are the ultimate responsibility of the stud groom.

Most mares will foal at night, usually between 10 p.m. and 3 a.m. Only a very few mares will foal during the day, but this does not mean that a constant watch should not be kept on all mares which are near to foaling. For this reason, during the day, mares which are near to foaling should be put in a paddock close to the stud buildings.

The person who is going to sit up usually arrives at 8 p.m. and stays until the stud groom comes out to feed the horses in the morning. It is usual for the stud groom himself to keep an eye on the mares between 5 o'clock when the staff finish work and 8 o'clock, when the person who is sitting up normally arrives. During the night any mares which are near to foaling should be looked at every 10–20 minutes and all other mares which are stabled should be seen at least twice during the night, in case they too start to foal or have an attack of colic, get cast, etc. All other animals which are inside should also be looked at during the night.

The equipment you will need for foaling should be ready at all times and ideally should consist of:

Tail bandage – for the mare;

Soap, water and a *clean* towel, for washing your hands thoroughly before foaling as well as afterwards;

A pair of very sharp surgical quality blunt-ended scissors – to cut open any stitched mares;

A jar of sterilising fluid (surgical spirit and Hibitane (ICI)) in which the scissors can be kept when not in use;

A quantity of antibiotic wound dressing powder – to treat the foal's navel immediately after foaling;

An oxygen cylinder, for use in emergencies;

Lengths of baler twine to tie up the after-birth (see plate 9).

A chamois leather and bucket of warm water or a dry clean towel – to help to rub the foal dry.

All foaling boxes must be equipped with electric light and should have heater lamps fitted, for use on early or sick foals. It is debatable whether it is better to leave the light on all night when a mare is near to foaling, or to let her get on with the job in the dark, if the light is left on all night she is not constantly disturbed every time the light is put on when the person who is sitting up looks at her.

Ideally the sitting-up room should adjoin the foaling box so that every movement of the mare can be heard, and a simple sliding window should be fitted in the wall, so that the mare can be observed at all times without being disturbed (see Fig. 5).

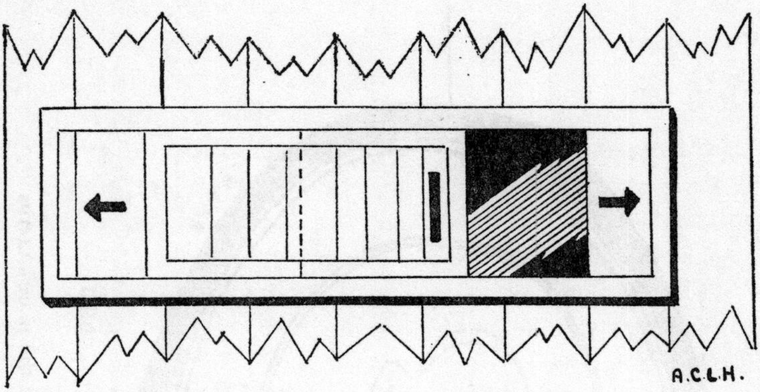

Fig. 5. Detail of inspection windows

The sitting-up room should be equipped with some form of heating, as it can get quite cold at night, even in late spring. There should be a sink with hot and cold water and most essential of all is an external telephone, so that the veterinary surgeon can be called easily in emergencies.

The whole process of foaling is divided into three distinct stages:

First stage: From the first onset of the labour pains to the rupture of the allantoic membranes (1st water bag);

Second stage: From the end of the first stage to the complete delivery of the foal;

Third stage: From the end of the second stage to the expulsion of the after-birth.

When the mare gets near to foaling she will often walk her box and paw at the floor. She may start to sweat profusely. She may get down to foal or commence foaling in the standing position, only going down as the foal begins to arrive – very few mares will foal standing. The first stage labour can take from as little as thirty minutes to as much as two hours or more.

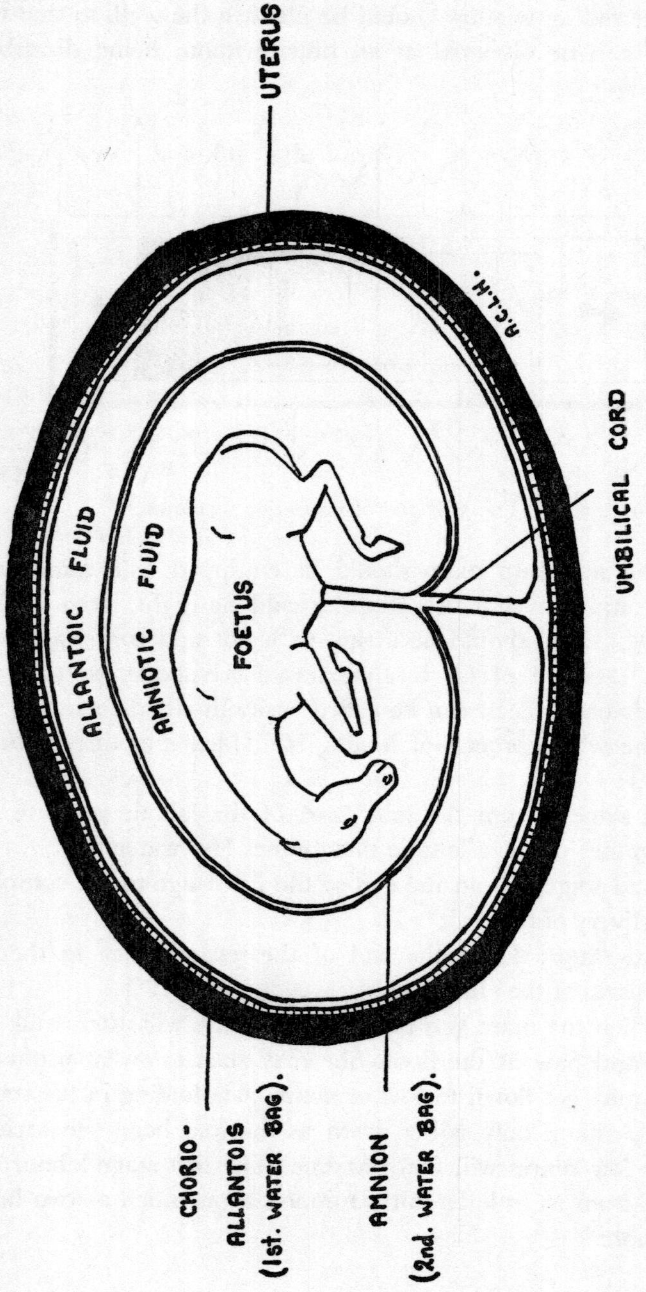

Fig. 6. Diagram of foetus and its membranes in the mare's uterus

The first sign of foaling is the breaking of the water bag, shown by a flow of yellow/brown water (allantoic fluid) from the vulva, which is the end of the first stage and marks the beginning of the second stage of labour. This is followed by the appearance of the amnion containing amniotic fluid (see plate 3). At this point, any stitched mares should be cut. The second stage usually takes from fifteen to fifty minutes.

If either the first or second stages of labour appear longer than normal, or the mare appears to be having difficulty foaling, it is better to call in your veterinary surgeon immediately, as the foal might be wrongly presented.

Soon after the first stage the feet and head will be observed inside the amnion. These should be checked to make sure that they are in the correct position (see Fig. 7). If there is

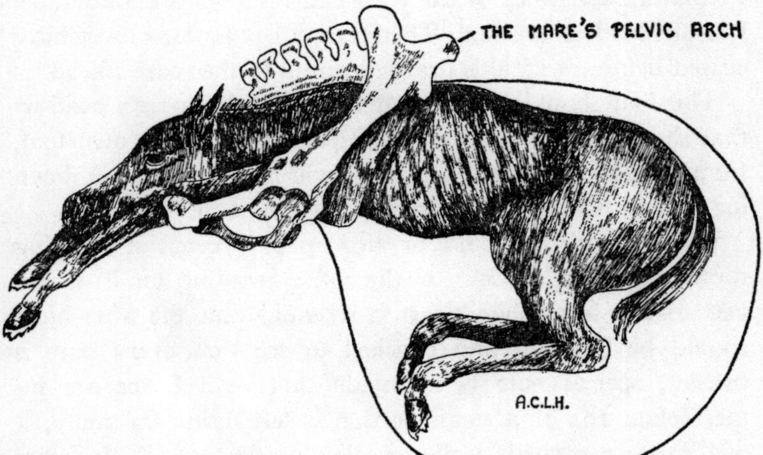

Fig. 7. Diagram showing a correct presentation of the foal

any major deviation from the normal, it is better to call in your veterinary surgeon *immediately*, rather than try to deal with the problem yourself.

At this stage, the mare may get to her feet only to go down again on her other side. This is thought to be a natural method of correcting a faulty position in the foal. The amnion may break as the front feet appear (plate 4) but if not, it should be broken by the attendant as soon as the head is out. If the amnion remains unbroken, the foal will suffocate when the cord severs.

Once the shoulders are out, the mare will often rest for some time before completing delivery (plate 5). When delivery is complete, it is probably a good idea to leave the foal's hind legs inside the vagina, as this will help to keep the mare down at least until the pulsations have ceased in the cord (plate 6).

Before birth, the foal receives its entire oxygen supply from the blood flowing through the cord from the mare (plate 7). When the cord breaks, the foal's breathing must be fully established as it then has to breath entirely on its own. Premature severance of the cord can result in a considerable loss of blood to the foal. Therefore, all pulsations must have ceased before the foal is dragged round to the mare's head.

The foal should be dragged round to the mare's head so that she can lick it dry and learn to recognise her own foal, from its taste and smell. The mare will probably remain down for at least half an hour after foaling (plate 8).

The after-birth should be tied-up to prevent it flapping round the mare's hocks or the mare treading on it as she gets to her feet (plate 9). It is essential that the after-birth should be laid out and checked to see that every part is present, special note being made that both horns are intact (plate 10). If a small portion is left inside the mare, it will cause a metritis (inflammation of the uterus). In cases where after-birth has not been expelled after 12 hours, your veterinary surgeon must be called to remove it, otherwise the mare may develop a metritis. In this respect, she is far more susceptible than the cow, which can be left for days without any apparent ill effects.

An accurate record should be kept of each foaling during the season and the details can be seen in Fig. 8.

Date	Name of Mare	Broke Water	Foaled	Cleansed	Foal Standing	Foal Sucked	First Meconium Passed	Sex	Colour	Weight

Fig. 8. Foaling record

This information will be of use to your veterinary surgeon should you need to call him later.

The owner should be notified at the first possible opportunity that his or her mare has foaled.

5

THE CARE AND MANAGEMENT OF YOUNG FOALS

As soon as the foal is born, it must start to adjust to its new way of life and environment. With the normal healthy foal this does not usually present any difficulty, but in the case of a sick foal veterinary help may be essential for its survival.

The new-born foal must have started to breathe by the time the cord is severed. Early severance of the cord will deprive the foal of a considerable quantity of blood and lead to anaemia or may even help to produce a 'barker' foal (see page 102). Normally, the cord will break about $1\frac{1}{2}$ inches from the umbilicus as the foal struggles to get up, or when the mare gets to her feet.

It can be broken by the attendant *once the pulsations have stopped* by dragging the foal round to the mare's head. This is a better method than cutting the cord, since it is less likely to lead to any infections, as the stretching and re-coil of the vessels produce a more natural seal. If bleeding from the stump occurs it is usually stopped, quite easily, by pinching the end between the fingers for a few minutes. The end of the stump should be liberally dusted with an antibiotic powder.

On many studs where mares are foaled inside, heat lamps are used to raise the temperature of the foaling box. However some authorities maintain that the sooner a foal hardens

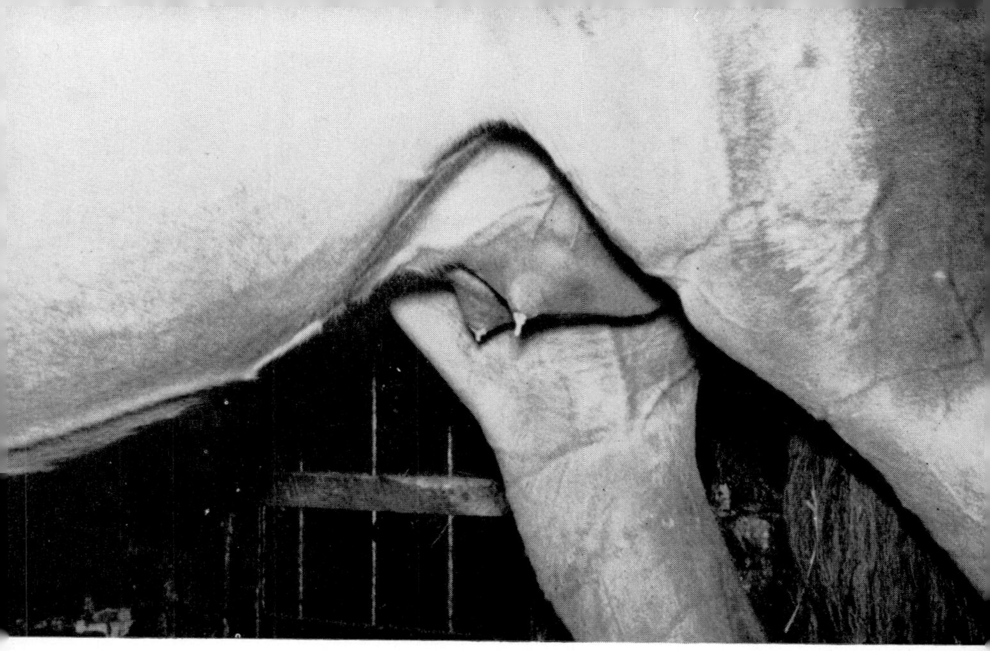

1. The mare's udder gradually gets larger during the last month of pregnancy and the muscles either side of her tail gradually get slacker. Candles of wax may appear on the ends of each teat. At this stage the mare could foal any moment. 2. In the case of some mares, the candles of wax disappear and they will run their milk from both teats before foaling

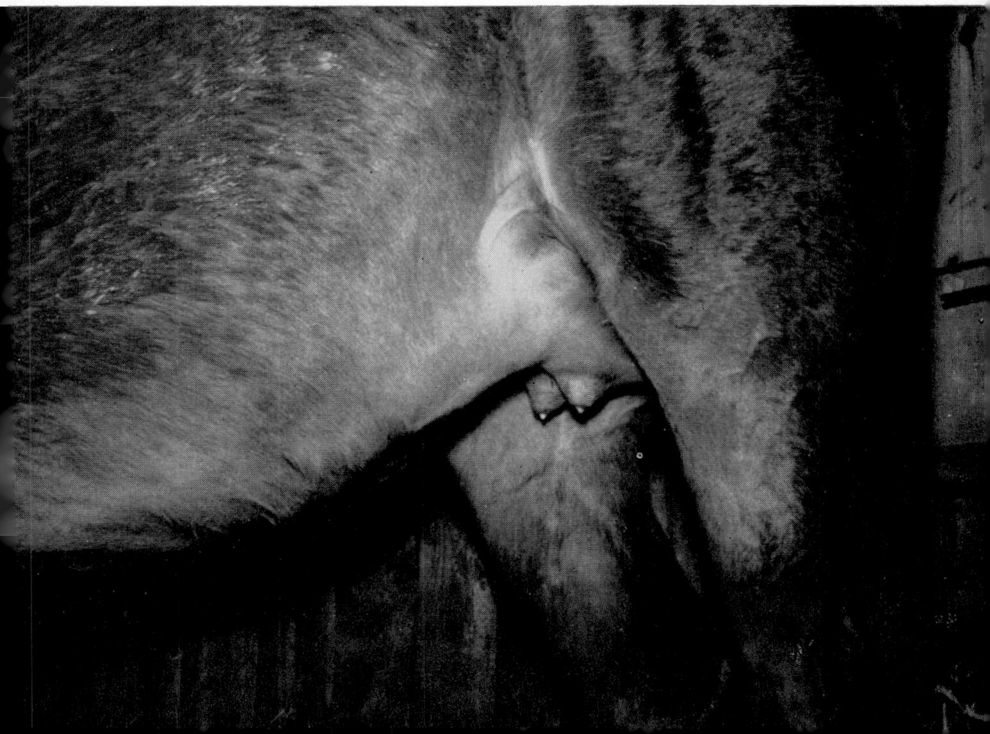

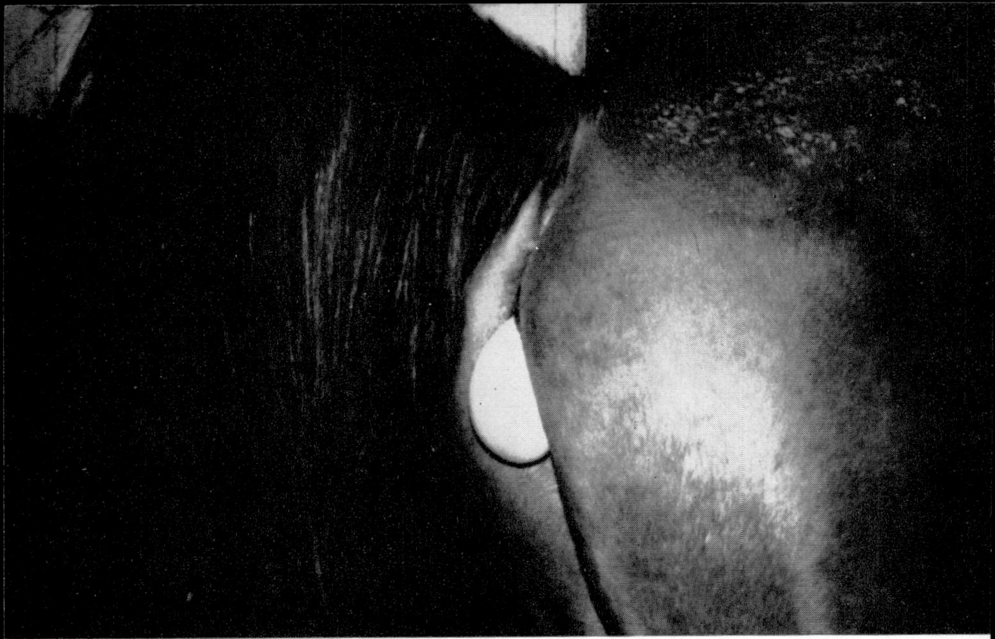

3. The first sign of foaling is the breaking of the water bag (placenta), shown by a flow of brown water (allantoic fluid) from the vulva. This is soon followed by the appearance of the amnion containing some amnotic fluid. 4. Soon afterwards the front feet should be seen inside the amnion. These must be checked to make sure that the presentation of the foal is correct. Should the attendant need to pull on the foal's legs, one must always be kept in advance of the other for this reason

5. Once the shoulders are out, the mare will often rest for some time before completing delivery. 6. Once delivery is complete, the mare and foal should be left to rest

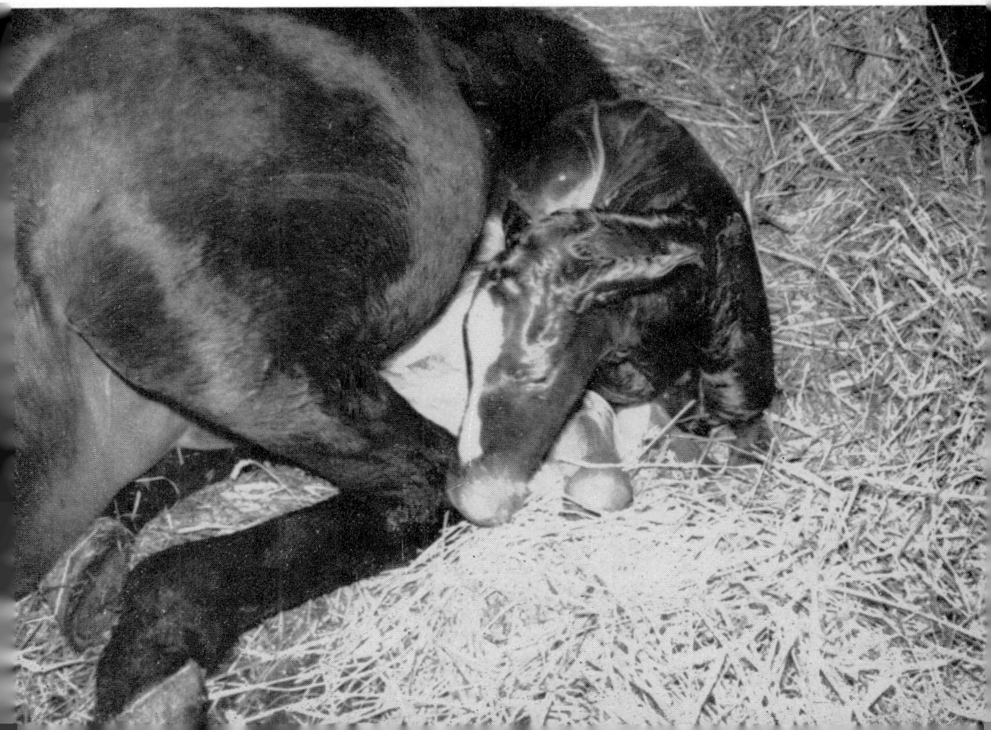

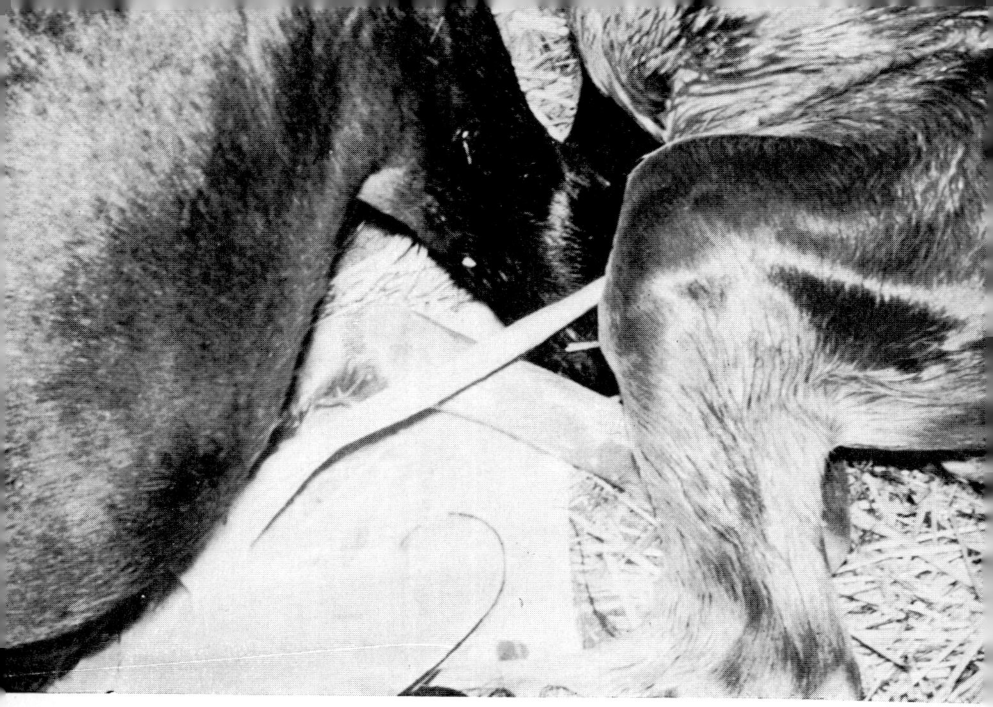

7. Before birth, the foal receives its entire oxygen supply from the blood flowing through the cord from the mare. 8. The foal should be dragged round to the mare's head so that she can lick it dry

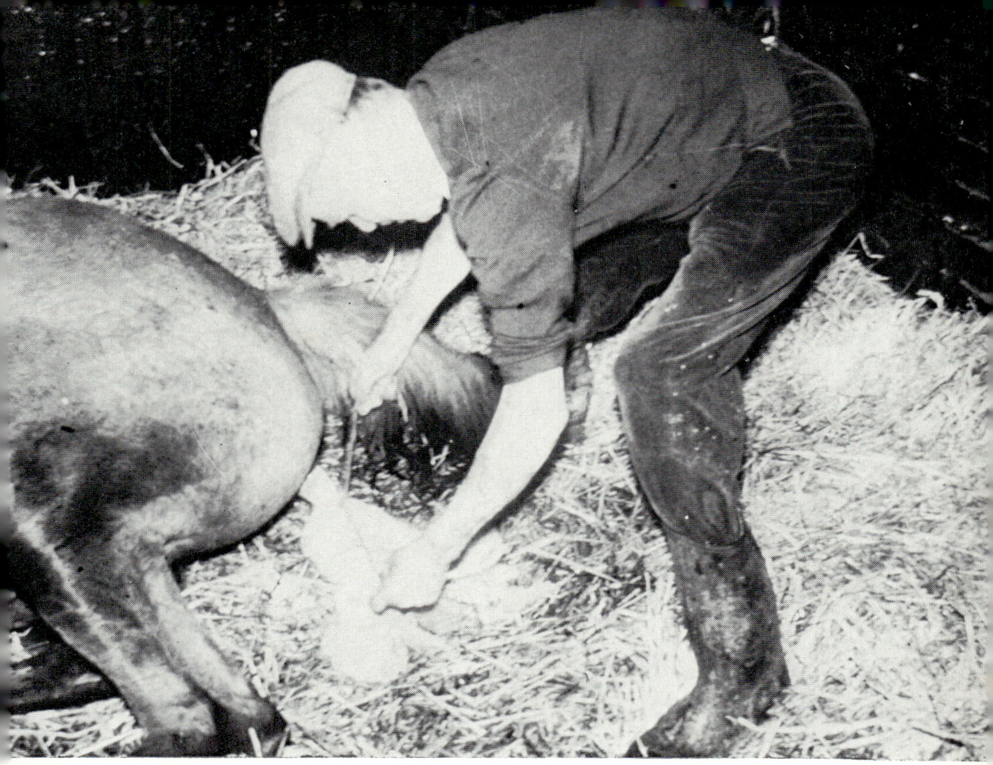

9. The after-birth should be tied or knotted up to prevent it flapping round the mare's hocks. 10. The after-birth or placenta. (1) main body of the chorio allantois (chorion) membranes (2) pregnant horn (3) non-pregnant horn (4) the amnion

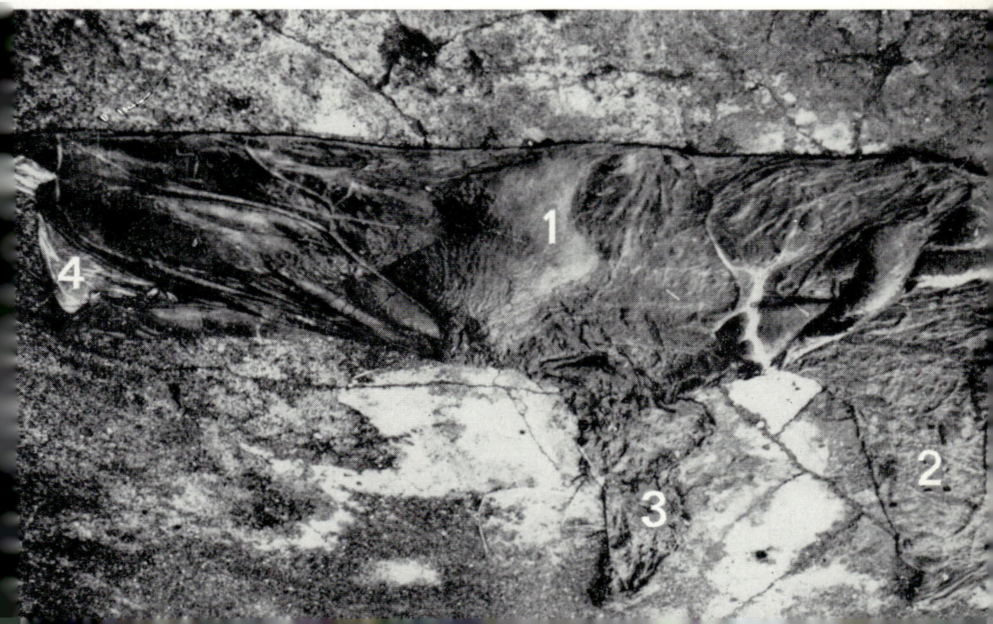

11. A trying bar built into the paddock fence. 12. The Arabian stallion Gerwazy, trying his mares over the paddock fence. Any mares which are in season will tend to go up to the stallion and show to him.

13. An ideal arrangement with cattle and mares grazing together. 14. Stocks – a good method of restraining mares for veterinary examination before covering

15. A conventional close-boarded stallion paddock. 16. The loose boxes should be completely washed down after the mares and foals have gone home

17. Wire mesh extension to a trailer prevents loose straw from blowing around the stud

off the better, and they never use artificial heating except for sick foals. It is better to foal all mares inside if the weather is cold and wet, as the sudden dramatic change in temperature from the womb, and lack of natural grease in the foal's coat, at this time, can lead to pneumonia.

If it is decided to foal a mare outside, great care must be taken to ensure that the paddock is absolutely safe, with the rails low enough to prevent a new-born foal from getting under the fence (see plate 12). There should be no streams, open ditches or ponds into which the foal can fall – as some mares tend to have the disastrous habit of foaling near water. When mares foal outside, there is always a risk that one may savage its own foal; or of another mare doing so before the foaling mare can get to her feet to protect her new-born foal.

As soon as delivery is complete, when the cord has severed and the mare can reach her foal, she will lick it all over – by doing this she is learning to recognise her own foal from the taste and smell of the fluids in which it is soaked. When the mare is foaled inside an attendant can assist in drying the foal off by rubbing it down with a *clean* chamois leather which has been rung out in warm water, or with a dry *clean* towel – in each case the cloth used should not smell strongly of other horses.

Before the foal is born, it is maintained in relatively sterile conditions, but at birth is suddenly comes in contact with every germ in its immediate surroundings. In order to survive, it must develop immunity to these organisms as quickly as possible. Unlike the human baby, the foal does not receive any resistance to disease before birth, but has to acquire all its immunity from the colostrum in the first twenty-four hours after birth. The level of antibodies in the colostrum can be specially raised against tetanus and influenza by vaccinating the mare one month before she is due to foal. Where new-born foals are highly 'at risk' to infection, as is the case on most public studs, it may be considered good policy to protect the foals at their most vulnerable time by antibiotic injection. By the time a foal shows obvious signs of illness, serious damage can have been done, as they have little resistance as compared with older horses. Prevention is much better than cure. If the mare is not vaccinated against tetanus, then tetanus antiserum should be given to the foal soon after birth. When dealing with valuable foals and where time permits, it is a good idea of take the temperature of every foal on the stud as a routine practice before turning out each morning. In this way, any slight rise in temperature can be noted and treatment given before serious infection becomes established.

The colostrum or fore-milk contains high levels of antibodies and the ability to absorb these across the small intestine ceases within thirty-six hours. Antibodies are substances found in the blood, which destroy germs that would, otherwise, cause disease.

After thirty-six hours the foal can only obtain immunity artificially, by means of injections or by natural development over a period of time. Where a mare has been running her milk before foaling, the antibody level will probably be low, so an alternative supply of colostrum should be found. It is a wise precaution to take a little colostrum after foaling from every mare which has not run her milk (as long as she has not produced a Haemolytic foal). Build up a colostrum bank in your own deep freeze – this will keep from the end of one season to the beginning of the next and this then available for emergencies.

Ideally, the foal should stand and suck within two hours of birth; in practice, the length of time taken varies enormously between individuals. Most foals show a marked suck reflex within five minutes of birth, and are drawn by areas of shade: they will tend to suck the walls of the loose-box or the mare's legs, before they are able to get up.

Foals will struggle to get up and fall many times before they finally manage to get to their feet. Periods of activity are usually divided by periods of rest before the foal finally gets control of its legs and stands. Unless the foal is weak, or the mare is being awkward, it is much better to leave the foal to get up on its own, making sure that there is plenty of straw on the floor of the loose-box, so that the foal does not injure itself.

Once up, the foal will be drawn by instinct to the darker areas under the mare's belly and between her legs. The mare can be encouraged to let her milk down by swabbing her udder with a piece of cotton wool soaked in hot water; this is a particular help where maiden mares are concerned. As long as the mare co-operates, the foal will gradually work towards the udder gaining strength and balance by the minute, until he finds a teat; this can take upwards of an hour.

Unless the foal is weak or the mare ticklish, it is unnecessary to offer any assistance to the foal. Physically forcing the foal's head under the mare and his mouth towards the teats will only confuse him and may delay the actual start of sucking. If a mare insists on following her foal round it is better just to hold the mare; if she attempts to kick the foal, it may be necessary to hold up one of her legs or take a handful of loose skin on her neck to restrain her, but *never put a twitch on a newly foaled mare*; this can cause the blood pressure to rise and could lead to haemorrhage into the ligaments of the uterus.

If your stud is to have a good name all foals must be returned to their owners quiet and easy to handle. Therefore, the foals must be handled and taught to lead as early as possible. A first-size foal headcollar can be put on the foal the day after it is born; this must be clean and soft to prevent it from rubbing the tender skin of the baby foal. If the weather is fine and warm, there is absolutely no reason why the mare and foal should not be let out into the paddock for a few hours the next day. It can be very dangerous to let young foals run loose round the stud behind their mothers; therefore some means of restraint should be used until the foal will lead quietly in-hand. The best method of leading very young foals is for two people to hold a sack round the quarters of the foal and link hands across its chest; in this way a young foal can be guided in complete safety anywhere with the mare, led by another person, behind. (see Fig. 9.)

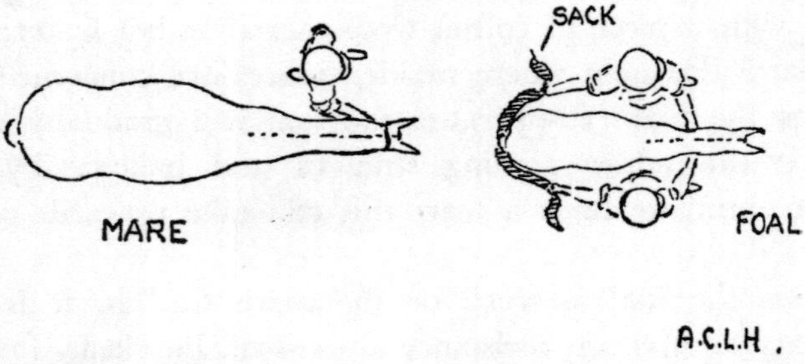

Fig. 9. How to guide a young foal without the use of a headcollar

The foal should be taught to lead from its headcollar as soon as possible; this is best done in the loose-box. When putting the headcollar on, make sure it fits reasonably tightly so that the foal cannot get a hoof through a strap, but not so tight that it will rub the foal. As foals grow very rapidly and soon outgrow their headcollar, someone on the stud should be made responsible for going round and checking all the foals' headcollars once a week to make sure that they still fit correctly.

For the initial lesson, slip a lead rein through the head-collar, get someone to lead the mare slowly round the loose-box and follow with the foal. The first time the foal feels a pull on the headcollar it will, almost certainly, run backwards and may rear or fling itself on to the ground, so a soft landing is essential. For this reason, never attempt to lead a foal, for the first time, in a concrete yard.

In order to encourage the foal to move forward freely,

place your right hand round the foal's hindquarters, keep your left hand on the rein and guide the foal round behind the mare. Once it gets the right idea, remove your right hand and hold the base of the foal's neck. You will be surprised how much control this gives you and what a steadying effect it has on the foal.

If a foal is lying down and you wish to get him up on to his feet, the easiest way of doing this is to straighten his front legs out and then run your hand down his back bone. In the majority of cases, you will find that the foal will immediately leap to his feet without any further trouble; but should you ever have occasion to lift the foal to his feet, never grab him with both hands under his stomach, as you may injure him. Rather, put one hand round his chest and the other round his hind-quarters and lift.

A large covered yard is a great asset on any public stud – in this way you can always give your mares and young foals some exercise, regardless of the weather. As previously stated, if the weather is fine and warm, there is absolutely no reason why a mare and foal should not be turned out the day after the foal is born.

Choose a small well-fenced and sheltered paddock. Get someone to lead the mare round for about ten minutes before you finally let her go; in this way the foal will get used to following its mother. Allow the mare to put down her head to graze and when she seems settled slip the lead rein. If, on the other hand, the weather is cold and wet and you cannot turn the mare and young foal out for a few days or you can only put them into a yard, it is essential – unless you know she is very quiet – to gallop the mare on her own in a small paddock, *before* you let her loose with her foal. Mares which have been confined to a loose-box, for any length of time, will tend to gallop off, kicking their heels in the air the first time that they are turned out. This is, of course, very dangerous; young foals have been killed by getting too close to the mare when she kicks out.

To gallop a mare, use a very well-fenced small paddock, get a sufficient number of people to stand round the outside of the fence, so that they can turn the mare as she gallops towards them. As soon as the mare is tired and stops galloping, she can be led back to her foal, which has been left – preferably with somebody – in the loose-box with the top door closed. Then both the mare and foal can be led out. The mare should be walked round and allowed to graze until she appears settled; then the lead rein can be slipped from her headcollar.

When turning mares and foals out into a paddock for the first few times, it is advisable to let them go singly, allowing each mare to settle before you turn the next one out. Otherwise you will run the risk of getting the foals mixed up, as young foals have a tendency to gallop off on their own, and will often run to other mares when they get lost. For this reason it is better not to turn out too many mares into a paddock together, until the foals are nearly a month old.

For ease of operation, a standard method of catching foals should be adopted throughout the stud and adhered to at all times. Any method which works can be used, but a well-tried and satisfactory method is as shown in Figs. 10 and 11. Eventually you will find that the foals will automatically take up the position shown, as soon as the mare is caught.

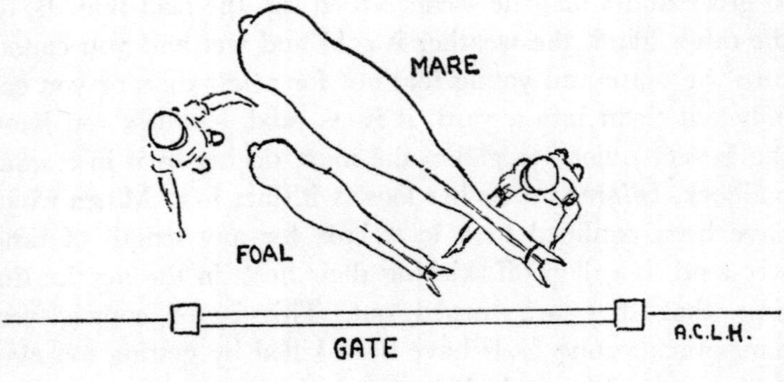

Fig. 10. How to catch a foal in a paddock

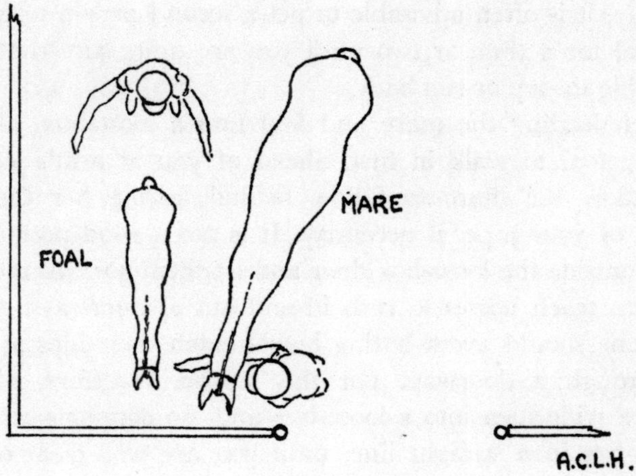

Fig. 11. How to catch a foal in a loose box

Initially the mare must be brought up to the gate in order to catch the foal; if you try to catch the foal near the fence, he may attempt to jump through, as you get hold of him.

When you first teach a foal to lead, it is advisable to take the mare and foal separately, getting one man to lead the mare in front, while another man follows up with the foal behind. As soon as the foal will lead really well from the head-collar alone, it can be led double with the mare as in Fig. 12.

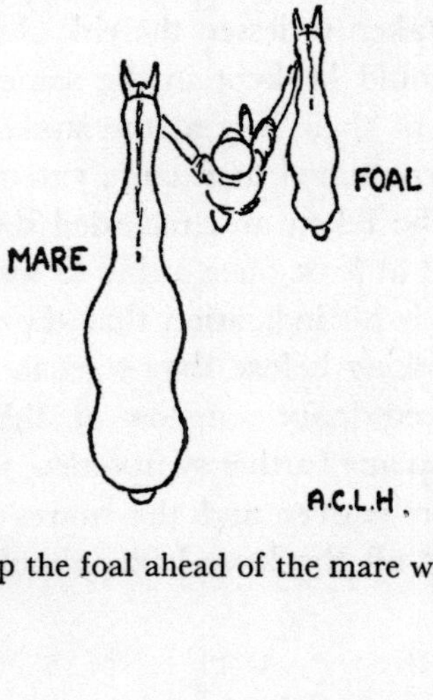

Fig. 12. Always keep the foal ahead of the mare when leading double

Initially, it is often advisable to get a second person to follow the foal for a time or two until you are quite sure that it is not going to stop or run back.

When leading the mare and foal into a loose-box, always get the foal to walk in first, ahead of you at arm's length, and allow the mare to follow behind, giving her the full length of your rope, if necessary. It is not a good practice to stand outside the loose-box door and let them go; this method tends to teach horses to rush in and out of doorways. At all costs one should avoid letting horses catch their hips as they go through a doorway. For this reason, therefore, always make a wide turn into a loose-box and, on coming out, walk the horses in a straight line, until you are well clear of the door-post before turning.

On no account should assistants be allowed to hit foals on their quarters as they let them go into the paddock or into the loose-box. When mares and foals are led out to the paddock separately, the person leading the foal must call out when he is ready to let the foal go; then, and not before, the person leading the mare should let her go. You will find yourself in trouble if the mare has already been loosed should the foal run back at the last minute.

One of the greatest problems on a public stud is scouring foals. There is no real method of preventing this all together, but steps can be taken to lessen the risk. As far as possible, a mare and foal should be kept in the same loose-box during the *whole period* of their stay at the stud: in this way, any infection can be contained to a large extent. The foals' temperatures should be taken and recorded daily and all mares should be checked at least once a day to see that their udders are empty which is an indication that the foal is well. Foals will often stop sucking before they start to scour. If they are treated by the veterinary surgeon at this stage they will usually not develop any further symptoms.

Once the season is over and the mares have gone home, it is essential that all the loose-boxes should be washed out, preferably with a power or steam washer and then re-painted or lime-washed (plate 16). Ideally the boxes should be left empty until the following season, which would prevent any build-up of disease, but in practice this is often not possible. In any case, all boxes should be thoroughly cleaned out between horses to prevent the spread of disease.

Very young foals should not be left out for long periods in wet and cold weather. Although it may not be raining at the time they could get a chill if they were to lie down on the damp grass. Once the foals get older and the weather improves, all the mares and foals should be turned out in the morning after breakfast and need not be brought in, until at least 4 p.m. or later, if your staff are willing.

Many Thoroughbred and Arabian Studs now leave their older foals out all night. This is the generally accepted form of management on most pony studs and has much to recommend it for other studs. It entails a great saving in labour (sometimes 45 per cent of stud expenses) and bedding and hay, which amounts to a sizeable sum by the end of the season. It is, however, essential that all foals are handled, and that the foals receive a feed each day. It is also vital that any Thoroughbred, or Thoroughbred-type, mares receive a supplementary ration so that they can milk well and also 'do' the foal they are carrying satisfactorily.

Under this system of management, it is advisable to get all the mares and foals in before mid-day, giving them a feed while everything is quiet before turning them out again. When they are brought in, it gives the stud groom an admirable opportunity to try each mare, examine them and check for any signs of injury or disease, which might be missed when animals are running out day and night in their paddocks. It also means that a regular worming programme can be carried out and any routine veterinary or blacksmith treatment given.

It would be inadvisable to bring in small pony mares for a feed, as this type of mare is often very prone to laminitis and is better left out on reasonably bare pasture. All the mares should have their manes and tails brushed through before they are turned out again each day.

When mares and foals are running out all the time, it is advisable to construct a creep-feed system in the paddocks, so that the foals can get some extra feed without competition from their dams. The usual method of doing this is to construct a square fence round some wooden feed troughs. The fence is made so that it consists of a top rail only, which is high enough to allow the foals to get under but not so high that the mares can follow them. Where mares of different sizes are visiting a particular horse, it then becomes necessary to put them in different paddocks according to their various sizes. The fencing used must be very strong to withstand being leant on by the mares. It is advisable to put strengtheners on the corner posts and put all the posts closer together than normal.

Most public studs make a routine practice of injecting all foals, at birth, with tetanus anti-serum and again one month later; this will protect the foal for the period of its stay at the stud. Once the foal is eight weeks old, it should be wormed and many studs make a habit of worming all their foals at this age which is often about the normal time when they go home. At five weeks, and/or before they go home, the foals should have their feet trimmed.

6

MARES, THEIR ANATOMY AND HORMONE CYCLES

Mares usually come in-season about seven to ten days after foaling although some, especially those with foals at foot, will show few, if any, external signs of oestrus. Ideally all mares should be checked for follicular activity by your veterinary surgeon at seven days after foaling, so that no mares are missed. He can then swab the mare and advise you on the general condition of the genital tract and say whether the mare is ready to be covered at the foaling heat or if it would be better to miss her and cover three weeks later, when she has had more time to recover. It was always considered the most fertile time to cover mares but recent research has shown that many mares which are covered at the foaling heat and prove to be in-foal, will either re-absorb their pregnancy or slip foal at a later date, due to not having been given enough time to clean up naturally after foaling.

Many studs have great difficulty in getting the services of a veterinary surgeon who is skilled in dealing with mares, so they prefer not to cover any of their mares at the foaling heat, but rather wait for them to come on again the next time. The other consideration is the high cost of veterinary treatment these days. Where mares are visiting low priced stallions, the value of any resulting foal would not justify high charges at any stage in its production. Therefore on this type of stud, you should be prepared to consult your veterinary surgeon, before actually calling him in, so that he can decide when to come and you can keep the charges to a minimum, without losing the efficiency of your stud.

Mares will usually come in-season every three weeks once their annual cycle has been established, and stay in-season for about five days, ie. they will return approximately every seventeen days from the time they go off.

At the beginning of the year mares will sometimes stay in-season for two weeks or longer without putting up a ripe egg, and this is particularly so in the case of maiden mares. There is little chance of getting a mare in foal until she is

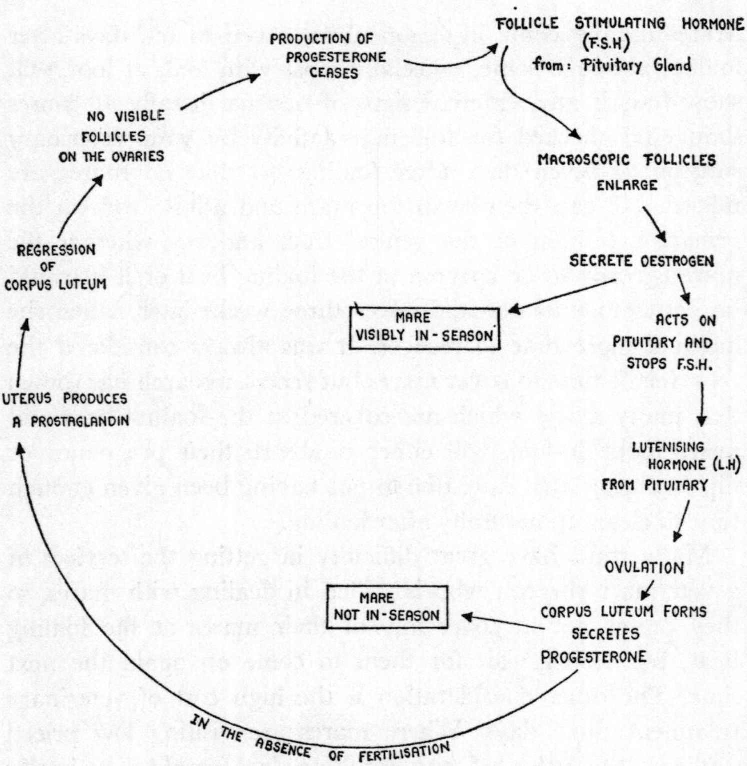

Fig. 13. Hormone cycle of a mare not in foal

functioning on a normal cycle which can be induced early in the season by leaving the lights on until last thing at night and by providing extra warmth with rugs or heat lamps. The hormone cycle of the mare which is not in-foal can be seen in Fig. 13. The external signs of sexual behaviour, that is alternate periods of acceptance and rejection of the stallion, are only a mirror of these internal physiological changes, as shown in the diagram.

A diagram of the mare's genital organs which also shows the swabbing technique, can be seen in Fig. 14. The mare

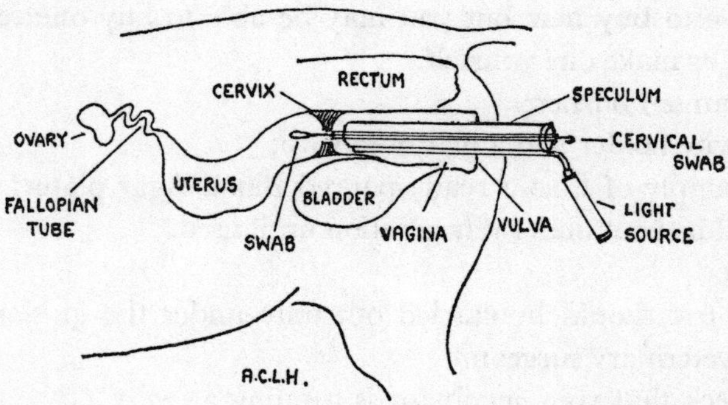

Fig. 14. Diagram of a mare's genital organs, showing the method of taking a cervical swab

has two ovaries, which are bean shaped and about the size of a small hen's egg when inactive. Each is suspended by a ligament from the abdominal cavity about an arm's length from the vulva.

Each ovary has several raised areas on its surface, one generally becoming much larger than the others. These elevations are evidence of the Graafian follicles which gradually ripen and in turn discharge a ripe egg (ovum), the process being known as ovulation.

The egg is discharged at the point on the ovary known as the ovulation fossa, which occurs in the middle of the concave side. As the follicle ripens so it becomes larger and softer, usually measuring from 2.5 cm to sometimes as much as 7.0 cm across before ripening. The egg itself only measures about 0.1 mm diameter.

During the days before ovulation the cervix gradually opens. *Once the cervix is open* the veterinary surgeon can take a swab for laboratory examination, to determine if the mare is clean and fit for covering. If your veterinary surgeon is agreeable, to save time and laboratory expenses this test may be carried out under his guidance on the stud.

Equipment needed

A laboratory incubator – to be run at 37 °C. These are expensive to buy new but you may be able to buy one second-hand, or make one yourself.

A bunsen burner;

A wire holder and a platinum wire;

A supply of Oxoid ready-poured Blood Agar plates, which should be kept under refrigeration until used.

Method

(This test should be carried out only under the guidance of your veterinary surgeon.)

Check that your incubator is running at 37 °C.

Take the number of plates you require – you can use one plate for two swabs if they are both put up at the same time. (See Fig. 15.)

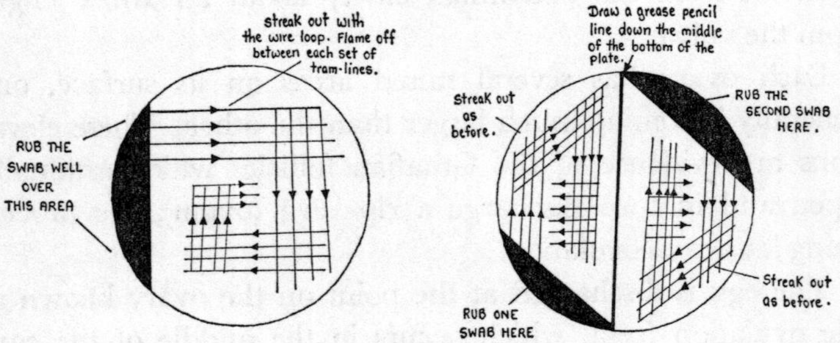

Fig. 15. How to plate out cervical swabs on to blood agar plates

Mark the plates on the bottom with the mare's name using a grease pencil. If you are putting two swabs up on the same plate draw a line down the middle of the plate, which will show through the agar, and place the mare's names on their separate halves of the plate.

Keep the plates upside down on the table.

Take the swab which corresponds to the name on the plate and rub it over an area on the plate which corresponds with Fig. 15.

Replace the plate on to its lid.

Take the platinum wire and bend its end into a small loop. Place the loop into your bunsen flame until it is very hot, in order to sterilise it.

Pick up your plate and test the wire round the uncontaminated edges of the agar until it is cool, so that you do not kill the organisms.

When cold, rub it well into the swabbed area and streak it out as shown in the diagram, flame and cool the wire between each change in direction.

Replace the plate on to its lid.

Incubate overnight in the upside down position.

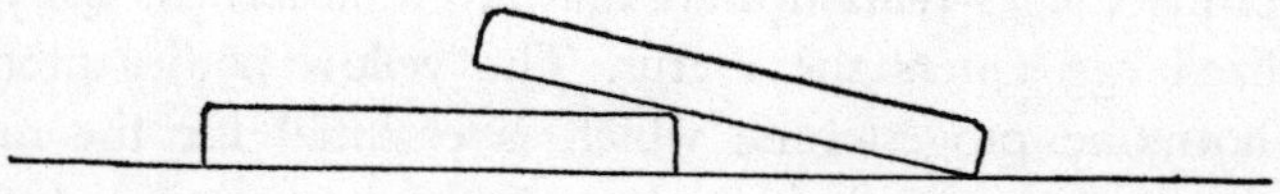

Fig. 16. How to dry blood agar plates in an incubator before use

By morning you will be able to see which mares are clean. If there are any organisms on the plates your veterinary surgeon will be able to tell you whether to cover the mare in question. If a mare is infected your veterinary surgeon will be able to send the plate away to a laboratory for identification and antibiotic sensitivity.

As already stated, on the majority of pony and hunter studs, it may not be sound economics to swab every mare before covering, although from the point of view of infection this is highly desirable. Generally speaking it is safe to leave all *genuine* maiden mares so long as you can guarantee that they really never have been covered before. Ideally all mares should be swabbed at the first heat after foaling, as many are found to be 'dirty' at this time.

Is *is* necessary to swab every barren mare before covering the first time. The reason for this is that there are some forms of low grade infection which do not always produce a vaginal discharge, but which are capable of being transmitted to the stallion and then to other clean mares. Also some forms of infection – e.g. Klebsiella – will cause sterility in both the stallion and his mares, and when an outbreak occurs, will cause a stud to be closed down for the rest of the season. Other more common forms of infection will cause abortion or sick foals at birth in the case of mares which happen to conceive. So it can be seen that swab testing of all barren mares and preferably all foaling mares, is an absolute must for well-run studs.

In the case of the mare which has been covered and is in-foal, the hormone cycle can be seen in Fig. 17.

Fertilization takes place in the fallopian tubes, and any un-fertilized eggs remain there and are re-absorbed. Only the fertilized egg enters the uterus. The yellow bodies produce the hormone progesterone which is essential for the maintenance of pregnancy in its early stages.

In the case of the mare which has failed to get in-foal, the uterus produces a hormone called prostaglandin which causes the yellow body to regress and therefore stops the production of progesterone. Prostaglandin, which has only recently been discovered, is considered to be a major break-through, since it causes mares to come in-season as and when required. However, there must be a high level of progesterone in the bloodstream first for it to work.

Prostaglandin's main use is to bring into season the type

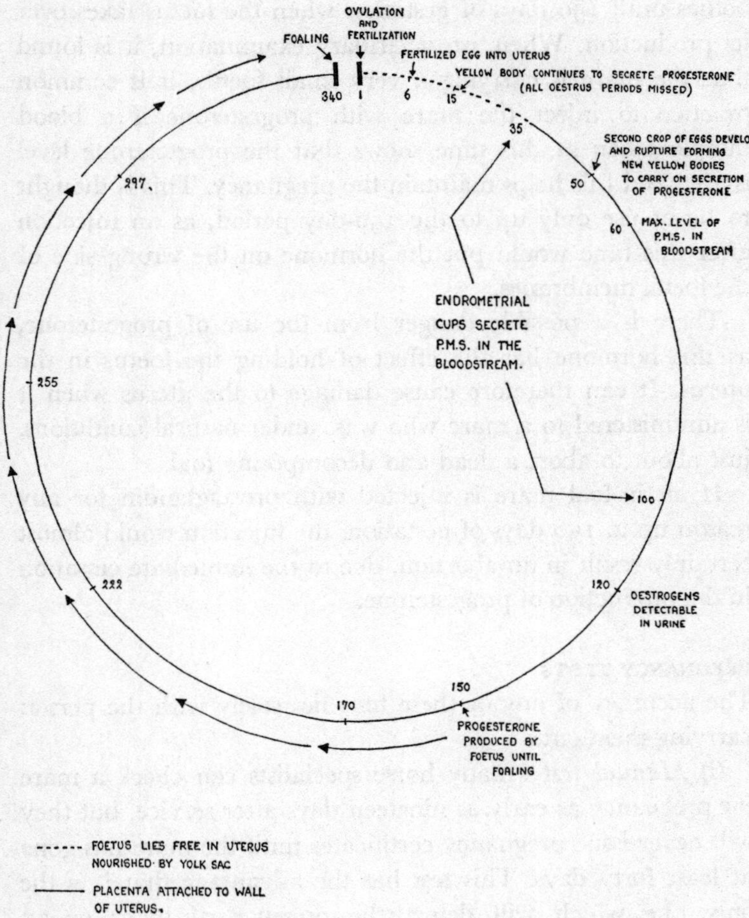

Fig. 17. Hormone cycle of a mare in foal

of mare which does not appear to come in for months on end (due to a persistant corpus luteum); or to bring a mare back into season when she has had twins washed out; or is reabsorbing a foetus; or not been covered due to being 'dirty', has then been treated and afterwards found to be 'clean'.

Progesterone on the other hand is produced by the yellow

bodies until 150 days of gestation when the foetus takes over its production. When, on veterinary examination, it is found that the mare is carrying a very small foetus, it is common practice to inject the mare with progesterone if a blood sample taken at this time shows that the progesterone level is low, and this helps maintain the pregnancy. This is thought to be of use only up to the 150-day period, as an injection after this time would put the hormone on the wrong side of the foetal membranes.

There is a possible danger from the use of progesterone, as this hormone has the effect of holding the foetus in the uterus. It can therefore cause damage to the uterus when it is administered to a mare who was, under natural conditions, just about to abort a dead and decomposing foal.

If an in-foal mare is injected with prostaglandin for any reason up to 150 days of gestation, the injection would almost certainly result in an abortion, due to the immediate cessation in the production of progesterone.

PREGNANCY TESTS

The accuracy of any of these tests lies solely with the person carrying them out.

(i) *Manual test* – many horse specialists can check a mare for pregnancy as early as nineteen days after service, but they will never issue pregnancy certificates until the mare has gone at least forty days. This test has the advantage that it is the only one which will detect the presence of twins or an undersized foetus which may be re-absorbed later and is best detected at thirty days.

(ii) *Blood test* – a blood sample (30 cc.) is collected from the jugular vein by a veterinary surgeon between fifty and ninety days after the last service date, and is tested in a laboratory, the optimium time for collection is seventy days.

(iii) *Urine test* – this is really outside the province of a public stud as it does not become operative until the mare has gone 120 days after her last service.

If desired a simple urine test can be carried out on the stud after a mare has gone 120 days. This test detects any Oestrogen in the urine sample, which will produce a green florescence, but the florescence is sometimes very difficult to detect, and can often only be seen with the aid of a torch in a dark room.

For the test:

Use a large pyrex test tube.

Mix:

10 ml. water,

1 ml. urine to be tested,

15 ml. conc. Sulphuric Acid.

A large amount of heat will be given off. Cool the tube immediately under a cold water tap and when cool, if possible stand the tube for about an hour in a refrigerator.

A distinct green colour should be seen in positive samples, negative samples remain brown, but as stated above this test is sometimes very difficult to read without a lot of experience.

Most mare owners like to have their mares checked for pregnancy around forty days before they take them home but it is a good idea to find out which test they prefer.

As soon as a mare has been tested in-foal the owner should be notified, many studs send a post card with the good news, including the last service date as a guide:

> Dear Sir/Madam,
>
> Your mare .... visiting .... was checked for pregnancy after . . . days and we are pleased to tell you that she has been examined IN-FOAL, to her last service ......
>
> Yours faithfully,

7

ANATOMY AND FUNCTION OF THE STALLION'S BREEDING ORGANS

For successful stallion management something should be known about the general reproductive anatomy of the male horse as well as that of the mare.

The most important male sexual organs are the testicles, which produce the sex cells and are responsible for male sexual behaviour and the masculine characteristics. The other organs are all concerned with the passage of the male sex cells from the testicles to the female genital tract. The testicles are oval structures about five inches long in the adult Thoroughbred horse, the left one often being slightly larger than the right.

During gestation, the testicles gradually move from a position near the kidneys finally descending through the inguinal ring and come to rest in the scrotum. In the equine foetus this descent is usually completed by nine to eleven months of gestation; therefore at birth two testicles should be clearly visible.

The scrotum is the pouch of hairless skin containing the testicles and situated between the thighs. Internally, it consists of muscle fibres which are responsible for the testicles being contracted when touched; fibrous tissue and a transparent membrane. In a few animals the testicles do not descend fully into the scrotum until the colt is two years old or sometimes

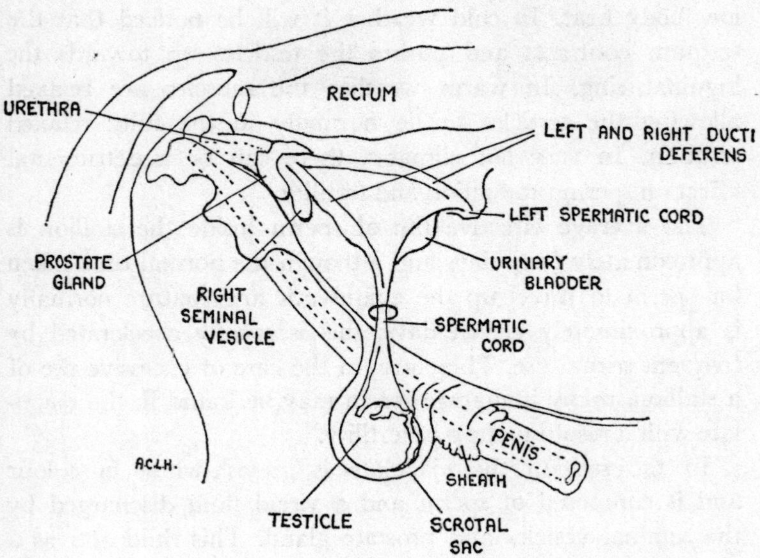

Fig. 18. The stallion's genital organs

even three or four years old, while more rarely still, they never descend at all. In these cases the animal is refered to as a rig or cryptorchid.

The primary functions of the testicles are the manufacture of sperm, which is a continuous process throughout the reproductive life of the individual and the production of testosterone – the hormone which is responsible for the male characteristics and sex drive.

The sperm are stored and matured in the ducts until required. The ducts originate in the testicles as minute coiled tubes in the lining of which the sperm are manufactured. During production, the number of chromosomes are reduced by half, the other half being ultimately supplied by the egg. The epididymal tubes (or ducts) are muscular and capable of contractions which propel the sperm through their length.

The scrotum and its internal structures form a temperature regulatory organ, because for maximum fertility the immature sperm should be maintained at a temperature just below body heat. In cold weather it will be noticed that the scrotum contracts and pushes the testicles up towards the inguinal ring. In warm weather the muscles are relaxed allowing the testicles to lie normally in the fully relaxed scrotum. In very hot climates, there can be a detrimental effect on sperm production and fertility.

The average effective life of sperm inside the stallion is approximately forty days and although the normal time taken for sperm to travel up the epididymis and mature normally is approximately eleven days, this is greatly accelerated by frequent sexual use. Therefore, in the case of excessive use of a stallion, many immature sperm may be found in the ejaculate with a resulting drop in fertility.

In the stallion the ejaculate is greyish-white in colour and is composed of sperm and a viscid fluid discharged by the seminal vesicles and prostate gland. This fluid acts as a transport medium for the sperm down the urethra and also helps to activate and nourish them at this time, as it contains electrolytes, citric acid, fructose, etc.

It has been estimated that each ejaculation measures approximately 60 mls. and contains approximately 200 million sperm per ml. As most stallions cover upwards of forty mares each season, and many of these considerably more than twice, and that there are six jets in some cases at each ejaculation, it can be seen that the testicles are very active organs and the excitement of service can make great demand on the energy of the horse during the stud season. With excessive services there can be a depression in sperm quality and unless a period of rest is allowed, any mares covered at this time will prove barren.

During service the sperm enter the uterus and in small numbers swim up the fallopian tubes where fertilization of the egg by one sperm will take place. Before fertilization can be effected an amount of debris must be removed from around the egg by an enzyme (hyaluronidase) which is secreted by the sperm.

The ability of sperm to fertilize an egg decreases rapidly with age, two days being considered the average length of time for most sperm, hence the traditional practice on studs of covering mares every other day until they stop being in season. However, under some unfavourable conditions, this time may be reduced considerably – e.g. poor sperm quality, infection in the mare or some incompatability between the mare's cells and the sperm.

The fewer numbers of sperm the semen contains the lower the fertility is likely to be, until a point is reached where there are insufficient numbers to fertilize the egg in any mare. There is, however, a marked seasonal variation in the quality of semen. Up to April there is usually a decrease in the amount of ejaculate and the number of sperm it contains, with a very marked increase during May and June. This is under the control of the pituitary gland in exactly the same way as in the mare and is influenced by the amount of sunlight and quality of the food available in exactly the same way. This point should be borne in mind when out of season coverings are contemplated. Better results possibly are obtained where the stallion's box is lit up by artificial light at night for a few extra hours in just the same way as his mares' boxes.

In the stallion follicle stimulating hormone activates the growth and formation of sperm in the testes and lutenising hormone causes the interstitial cells to liberate the hormone testosterone which promotes the male characteristics and sex drive.

When it is decided to buy a colt for stud purposes, it is a good idea to have him examined by a veterinary surgeon before the sale is completed, and afterwards take out an infertility insurance. The insurance gives a 100 per cent refund in the case of complete infertility. (see Chapter 17.)

When a veterinary surgeon examines a colt as a prospective stallion, he will pay particular attention to his heart. As can be seen, stud duties put a strain on the horse and each year a few horses suffer from heart attacks as a direct result of covering mares. Lack of fitness and excessive fat in stallions are predisposing causes of heart failure. Your veterinary surgeon will also examine the external genital organs; both tecticles should be visible and of normal size. Where one testicle is retained the animal will probably still be fertile but

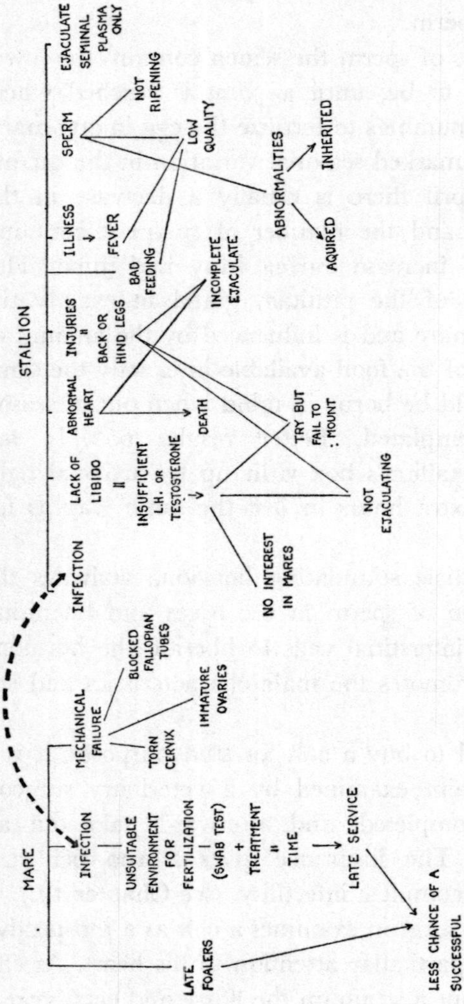

Fig. 19

Factors affecting fertility in stallions. In most cases an improvement in fertility percentage is shown when L.H. is used to sychronise ovulation in the mare with service by the stallion; in theory one can expect 100 per cent fertility with 40 to 50 mares per season, in practise at least 20 per cent of the mares covered will be ba rren each year (the national foaling average for Thoroughbred mares is 1 year barren for every 3 years at stud). 70 per cent fertility is therefore regarded as quite satisfactory. 60 per cent or less can indicate a low fertility on the part of the stallion he cannot be passed as sound for stud purposes. The veterinary surgeon will also look for any unsoundness in the hind limbs and back, as service of a mare puts considerable strain on these regions and any unsoundness here can be the direct cause of a horse not covering his mares successfully. Unsoundness in the fore legs due to injury or natural wear from hard work is not so important in a stallion.

Later, when the colt covers his first mare or a special test mare, a sample of his semen can be collected for laboratory examination. In fact, if the semen is collected into an artificial vagina, this gives a much better sample for examination.

The quality of a semen sample is measured by the number of sperm present, the number of live sperm per ml. of sample, their morphology; and motility; and the percentage of abnormal sperm is recorded. These are acquired or inherited defects in the head or tail of the sperm and an average normal sample may contain approximately 16 per cent of abnormal sperm. It should be noted, however, that the first service of the season will contain more abnormal sperm than later coverings.

8

TRYING MARES

Trying is the process whereby man determines the sexual state of a mare at any given time, using a male horse for this purpose.

Under natural conditions where the stallion is running with his mares, the horse himself will determine exactly when the mares are in-season and ready for service.

When the stallion and his mares are kept apart it is entirely up to the stud groom to determine the optimum time for successful service. Therefore trying is a major factor influencing the fertility percentage of any horse. Where the stallion has a full book of mares, it is common pactice to cut down the number of services any mare has to a minimum, which means covering within forty-eight hours of ovulation, to give the best possible chance of conception.

The external signs of sexual behaviour are merely a mirror of the physiological changes which take place inside the mare but unfortunately in a few cases, they are not always an accurate guide. In these instances, the services of a veterinary surgeon (is essential), to determine the exact stage the mare has reached in her cycle.

Where the stallion and his mares are kept apart, except for the periods of trying and covering, it is essential to find out exactly when the mare is in-season. Normally mares remain in-season for five to seven days and are off for seventeen days, which gives on average a twenty-one-day cycle (see page 60). All mares on the stud should be tried every other day. This does not include recently covered mares which can be left for ten days from the last day they were in-season and mares which have been tested in-foal, and foaling mares for the first six days after foaling.

There are various ways of trying mares and any which are both effective and safe can be used. Broadly speaking the ways one can discover if a particular mare is in-season are:

(i) Lead the stallion up to one side of a trying bar and lead the mare up on the other side.

(ii) Lead the stallion up to a trying bar constructed in the paddock fence (see plate 111). Allow the mares which are out in the paddock to approach the horse.

(iii) By quiet observation of the mares in the paddock.

(iv) Lead the stallion among the mares running lose in the paddock.

(v) Confine the stallion in an enclosure near to the mares' paddock.

(vi) Turn a very small stallion into a little paddock for a short time with a few tall mares, and keep them under constant observation.

(vii) Exercise the stallion by walking him along the outside of the mares' paddock.

(viii) Try over the loose-box door.

(ix) Turn a vasectomised stallion with the mares; or as an alternative to (ix)

(x) Retroversion of penis – surgery to change the direction of the penis so he cannot penetrate the mare.

(i) THE STALLION AND MARE LED IN-HAND TO OPPOSITE SIDES OF THE TRYING BAR

This is the most common method of trying mares which are not actually running out with the stallion. Diagrams of the trying bar and the various layouts can be seen in Figs. 20 and 21.

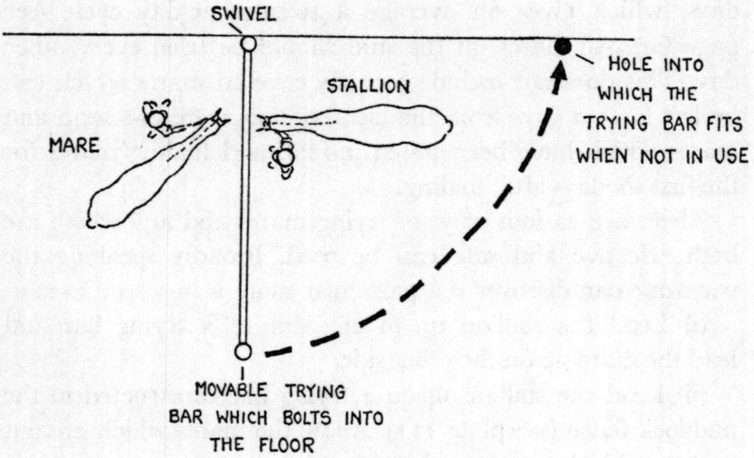

Fig. 20. A trying bar with moveable bar

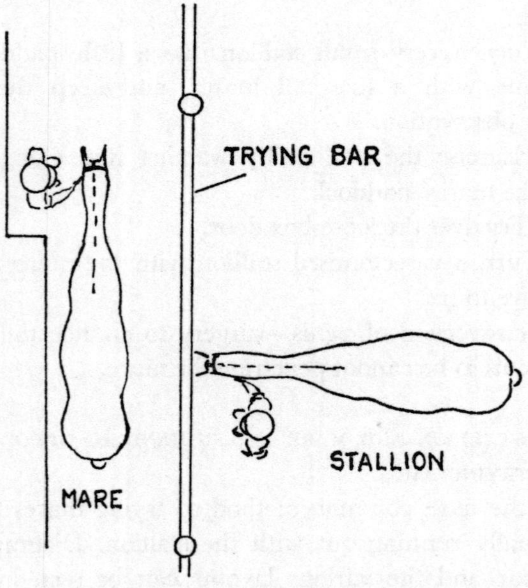

Fig. 21. A 'crush type' trying bar

For trying, both the stallion and mare must wear bridles; this is particularly important as far as the stallion is concerned, as should he become over-keen or excited he might well try to jump the bar to get to the mare and with only a headcollar on, it would be virtually impossible to stop him.

Trying mares does present some slight danger for the stallion; therefore, on large studs or where the stallion in question is very valuable it is common practice to use an inferior male horse to do all the teasing and so save the stallion for covering only. This type of male horse is known as a 'teaser'.

When a 'teaser' is employed he should be allowed to cover at least one or two mares in a season in order to maintain his interest in trying mares. An ideal 'teaser' will talk to his mares quietly and will not lose interest half way through the operation. A noisy teaser who screams and shouts and bites the mares will only frighten the shy ones. However, a teaser should make sufficient noise to call the mares up when he is trying out in the paddock.

Trying bars should be sufficiently high (usually about 4 ft.) to allow the stallion only to get his head and neck over, the rest of him remaining protected from kicks. When a very small stallion is used the ground should be built up on his side, so that the bar is still too high on the mares' side for even the tallest mare to kick over the top. The bar should be 8 ft. long so that a mare is unlikely to kick round it.

To prevent the mares from injuring themselves when they kick out, the bar should be heavily padded with coconut matting or some similar substance (see Fig. 22). The supports should be metal or very heavy timber set firmly in concrete and the bar itself should be made of half wooden sleepers or boards of a similar thickness – boarded up and down, rather than across, which makes for easier renewal of any broken boards. A free running, revolving pole can be fitted to the top as an extra safety precaution, in case the stallion ever tries to

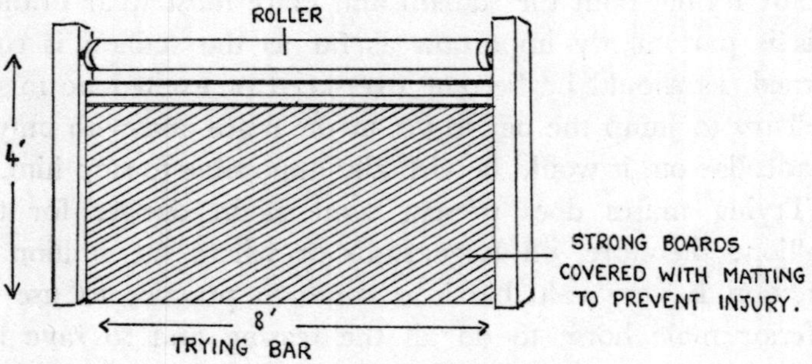

Fig. 22. A covered trying bar

jump over and this will help to make pulling him back easier.

Always make certain that the trying bar is located so that the stallion does not stand with his back to a wall, as he might break a bone in his hoof or pastern, should he become excited and lash out during trying.

The stallion should be held back slightly from the bar as the mare is led up; this makes it easier to persuade the shy mares to approach the horse. The stallion and mare should be presented head on so that they can talk to one another. Someone responsible should stand at a safe distance behind the mare so that he can see if she is in-season; he can then record each mare's behaviour in the trying book. The man holding the mare must make certain that he has a long enough rein and stand up by her shoulder all the time, as some mares will strike out with their front feet during trying or even rear and strike out, so if he is standing in front of the mare he is very likely to get hurt.

Once the mare and stallion have greeted one another, the mare will usually swing round with her hind quarters towards the boards. If she is not in-season she will usually lay back her ears, squeal and kick. If, on the other hand, she is in-season she will tend to lean towards the bar, stand with her hind legs apart, lift her tail and pass water, usually showing 'winking' of the clitoris.

The expression on the mare's face as she approaches the stallion in the first place will usually tell you, when you know the mare, if she is in or out of season. Constant and prolonged teasing of mares which are not in-season will often lead to abnormal behaviour at the trying bar. These mares will tend to pass water even though they are not in-season, and very often the only difference between them and those genuinely in-season is that they will lay back their ears and swish their tails violently from side to side. It is therefore not a good idea to continue trying a mare for a long time on any one day, unless you are in genuine doubt about her sexual state.

Mares which will not show readily to the horse should be pushed over towards the bar so that the stallion can tease them further.

This is the best method of trying mares with foals at foot, in which case the foal is left in the loose-box with the top door closed with someone to keep an eye on it, while the mare is lead to the bar for trying.

(ii) LEAD THE STALLION UP TO A TRYING BAR CONSTRUCTED IN THE PADDOCK FENCE

A photograph of this type of trying bar can be seen in plate 11. The one shown in the photograph is of the permament type but an equally effective form of bar can be made by hanging some boards nailed together to form a solid bar on the paddock gate (see Fig. 23).

The stallion is brought to the trying bar in the paddock fence, and a good teaser will call up his mares and those which are in-season will usually answer his cries and come running up to greet him. Otherwise the mares should be driven up the paddock towards the bar. Any shy mares which hang back and will not approach the horse, should be caught and led up to the bar for trying. It will be observed that there are always some aggressive mares in each paddock which will rush up to the teaser and remain at the bar kicking

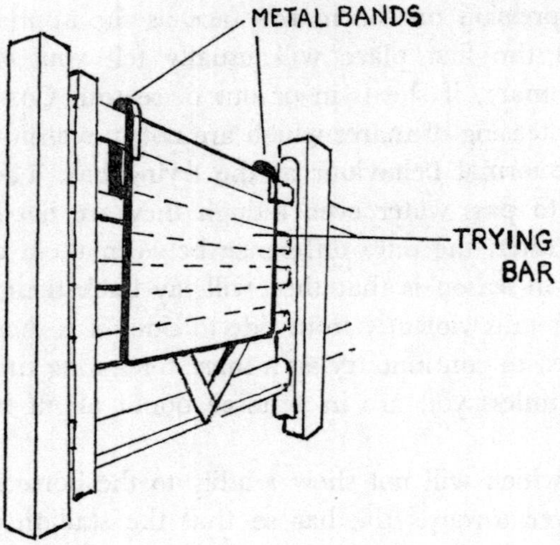

Fig. 23. A trying bar hung on a paddock gate

and squealing and preventing any other mare from approaching the horse. Once they have been tried, this type of mare should be driven away.

As with other methods of trying, you should stand at a safe distance behind the mares in the paddock to observe which, if any, are in-season.

Unless very great care is taken, this is an unsuitable method of trying mares with foals at foot, since during the trying process there is always the danger that a foal might get kicked. The only really safe way to try this type of mare is to catch up each one and try her individually as already described.

(iii) BY QUIET OBSERVATION OF THE MARES IN THE PADDOCK

It is a very good practice to wander round the paddocks each evening before dark. Unless your dog is very well behaved, do not take him with you as he will only tend to disturb the mares. At this time of day it is often possible to spot a shy mare in-season. These mares will often show to other mares in the paddock when they will not show to the teaser during the day; and the best time to observe them is when everything is quiet and they are undisturbed and grazing peacefully.

Any mare which shows to other mares but not to the teaser can be brought in for examination by the veterinary surgeon, who will be able to tell you if she has a ripe egg and is ready for covering. This type of mare will often relax and show well to the stallion at the time of covering.

Any mare which has been at the stud for longer than three weeks without showing signs of coming in-season should be caught up for the veterinary surgeon to examine and for possible treatment.

(iv) LEAD THE STALLION AMONG THE MARES RUNNING LOOSE IN THE PADDOCK

The stallion must wear a bridle and should be led by a lungeing rein attached to the bit. The man leading the horse should also carry a long whip, so that he can protect himself and the stallion from any aggressive mares.

This is an unsuitable method of trying mares with foals at foot as there is a great risk that some of the foals might get kicked by other mares or by an over-excited stallion. This can also be a risky method of trying where a teaser is used, as should he happen to get loose some accidental coverings might result!

Where there is a labour shortage on the stud and the stallion is not so valuable that it matters too much if he gets kicked, the horse himself can be led among the mares and any that are found to be in-season can be covered there and then. This is not, of course, an ideal method as it does not allow for any infected mares being treated before service nor does it offer much safety for the stallion, who could easily get kicked particularly as he dismounts. To reduce the risk slightly, a second man should accompany the person leading the stallion, then he can hold any in-season mare while she is being covered and so help to lessen the chance of injury to the stallion.

(v) CONFINE THE STALLION IN AN ENCLOSURE NEAR TO THE MARES

The stallion, in this case, is not usually the horse which is actually used for covering the mares, as he would take too much out of himself worrying about the mares all day long.

The fence between the stallion and the mares must be both high and extra strong for obvious reasons, but the stallion and mares must be able to see one another easily through it. Any mares which are in-season will be observed grazing near to the stallion's fence and taking an obvious interest in the horse. They can then be caught up and tried properly. This method is safe for mares with foals at foot but is not a system commonly used in the British Isles.

(vi) TURN A VERY SMALL STALLION OUT IN A LITTLE PADDOCK WITH LARGE MARES

This method is also not commonly seen in the British Isles. It is only of use where the mares in question are of the Thoroughbred or hunter type, standing at least 15 h.h. and the stallion is a very small Shetland type, measuring inches rather than hands.

The mares to be tried are placed in a small paddock, yard or corral so that it is easy to catch the stallion again after trying. The pony could be left with the mares in their normal paddock all the time but there is a slight risk of a clever pony covering a mare while she is lying down, or in the case of a hilly paddock getting her on a suitable incline. Therefore it is essential to use a smaller flat area and observe the pony and the mares the whole time he is with them.

(vii) EXERCISE THE STALLION BY WALKING HIM ALONG THE OUTSIDE OF THE MARES' PADDOCK

A photograph of this method of trying mares can be seen in plate 12.

This method is only suitable for mares which have not got foals at foot, as there is a great danger of the foals getting kicked when the mares come crowding up to see the horse. It is, however, possible to try mares this way with foals at foot if only a few mares are turned out together. By using this method, mares can be tried every day without any extra labour requirement.

Ideally, someone should stand inside the paddock so that he or she can observe which mares are in-season and can make a note of them in the trying book.

Great care should be taken, when using this method, that the stallion does not get over-excited and put a front leg over the fence or that a mare swings round and puts her leg through a rail.

(viii) TRY OVER THE LOOSE-BOX DOOR

This method is really only to be recommended where there is a shortage of labour on the stud, as although it can be carried out quite satisfactorily by one person on his own, it does tend to have a rather detrimental effect on the doors and immediately surrounding woodwork of the loose-boxes!

Either the stallion can be led round from box to box, trying the mares over their top doors, but great care should be taken in the case of all mares with foals at foot; or the mares can be led to the stallion's loose-box so that he can try the mares over his door. This way is probably the more satisfactory as it is easier to observe if the mare is in-season when she is outside than it is when she is in her own loose-box; also there is no danger to the foal.

When the mares are tried outside the stallion's loose-box, the area around the door must be well flooded with a strong smelling disinfectant afterwards, otherwise the horse will start worrying if he can smell in-season mares all the time and will soon lose condition.

(ix) TURN A VASECTOMISED STALLION WITH THE MARES

A horse which has been operated on in this way will take a normal masculine interest in his mares. He will be able to mount and enter the mares, but will be unable to serve them.

The attentions of this type of horse will help to bring the mares in-season and he will also get any maiden mares quiet for service, which will eliminate any necessity for 'bouncing' by the teaser before service by the stallion. This method does, however, lay the stud open to a general spread of vaginal or uterine infection, via the vasectomised stallion.

(x) TURN A STALLION WHICH HAS BEEN OPERATED ON TO PRODUCE A RETROVERSION OF THE PENIS WITH THE MARES

This has the same effect as ix, but eliminates any chance of spreading infection, since although the horse can mount his mares, he is unable to enter them.

All trying of mares should take place first thing in the morning so that you can decide which mares can be covered that day and plan your time accordingly, also you will then know which mares need to be examined by your veterinary surgeon.

When two mares are to be covered, one should go to the horse about 9 o'clock in the morning and the other about 3 o'clock in the afternoon. If three mares are to be covered, the first should visit the horse at about 7.30 a.m., the second at around 12.30 p.m. and the third at 5.30 p.m., which allows a minimum of five hours' rest between coverings for the horse and should ensure that each mare has an equal chance of getting in foal.

9

COVERING MARES

As in the case of trying mares, both the stallion and mare should wear bridles for covering so that the handlers have the maximum control over both at all times.

Ideally mares should be covered in an indoor yard with a sand floor which is kept damp to reduce the chance of dust-borne infections. The sand also gives an ideal non-slip surface for the horse. Concrete or Tarmac are highly unsuitable surfaces on which to cover mares as the stallion could easily slip. Grass is quite suitable as long as it is not so muddy that the horse gets dirt on his penis before covering.

The actual place chosen must be out of sight of your neighbours otherwise you may receive complaints, and rightly so! This is particularly important when you are travelling your stallion to visit mares at home and you have to improvise arrangements for trying and covering.

In the case of mares with foals at foot, the foal should be left in a loose-box with someone to look after it until the mare returns, or left un-attended with the top door closed, so that it cannot try to jump out. Alternatively, some people prefer to take the foal to the covering yard or area where covering takes place and hold it out of harm's way in front of the mare, so that she can see it at all times. This method does in fact seem to settle some mares, who would otherwise be worrying about their foals, while being covered. Covering should always be as quiet a process as possible with the mare completely calm and relaxed, although this is not always too easy to achieve. Anything which will help to obtain it, is well worth trying. Bring the mare to your selected covering site with a bridle on, have a twitch ready but do not use it unless it is absolutely necessary. Most mares are more relaxed without a twitch, and some are even more awkward with one. Put the covering boots on the mare's hind feet – these are preferable to hobbles which can upset a highly strung mare (plate 18). Covering boots are usually made from heavy duty felt and can be obtained from most saddlers.

A tail bandage should be put on the mare to keep the hairs out of the way (plate 19); pieces of tape tied round the dock at intervals of about 2–3 inches can also be used, or a long piece of string can be laced down the whole length of the tail and the loose end of the string used to hold the tail out of the way. Whichever method is used, the important thing is to get all the hairs at the top of the tail out of the way of the stallion's penis at the time of covering, since a hair entering the vagina could cause a small painful cut. At this time blinkers (racing or otherwise) can, with advantage, be put on any young or un-broken fillies, which might otherwise take fright when they see the stallion jump.

While covering, some stallions will tend to bite their mares, in which case a leather neck guard should be used at each covering. A leather or rope strap fastened round the mare's neck for the stallion to get hold of will, in some cases, help a small stallion when he is covering a large mare.

The mare's vulva, hindquarters and dock region should be well washed down with a mild antiseptic solution which is not strong smelling (plate 20).

As soon as the mare is ready for covering, the stallion can be brought out to her. He should be allowed to approach his mare on her near side, so that he is out of the way should she kick (plate 23). Some authorities prefer to hold the horse back from the mare until he is absolutely ready to cover, then bring him forward and let him jump immediately, claiming that he is less likely to get kicked this way. It is usually preferable to let the horse try his mare for a minute or two before mounting. Many mares which are fully in-season will pass water as the stallion approaches and stand with their hind legs well apart.

Keep the stallion on the near-side of the mare out of harm's way, allow him to tease the mare until he is fully drawn, his back muscles are tense and he is absolutely ready to cover. On no account allow him to jump until he is really ready.

If possible have an extra man on the off-side of the mare to hold her tail out of the stallion's way (plate 22). If only two people are available, the person handling the stallion should pull the mare's tail out of the way of the horse as he mounts.

When he is ready to cover, the experienced stallion will automatically swing himself round behind the mare as he jumps (plate 24). The novice horse covering a quiet mare, should be kept directly behind the mare so that he is positioned correctly for covering. As the stallion mounts his mare, her natural reaction is to move forward a step or two, this should be allowed but she should not be permitted to walk any further (plate 26).

As the stallion ejaculates, so his tail will move up and down – known as 'flagging'. Note the positions of the stallion's tail, in plates 26 and 27. 'Flagging' is most noticeable in Arabian stallions and others with a high tail carriage. Ejaculation can be felt by placing your hand lightly under the stallion's penis as he serves his mare. This is the only infallible way of telling if the stallion has actually covered his mare.

It sometimes helps to steady the horse if the stallion man holds his front leg while he covers the mare (plate 25), and this is particularly useful in the case of the small stallion covering a large mare. Where there is a great discrepancy between the size of the stallion and his mare a mound can be made for him to stand on and the mare backed up to the base. She must, however, be held quite still when the horse jumps otherwise the stallion will finish up on the same level again. If, on the other hand, you dig a hole for the mare's hind legs, she must also be held quite still otherwise the stallion will finish up in the hole himself. The reverse applies when a large stallion is covering a small mare.

As the stallion comes off the mare, her head should be turned to the left, this will have the effect of swinging her hind-quarters to the right, away from the horse, which is essential should she lash out (see Fig. 24 and plate 30).

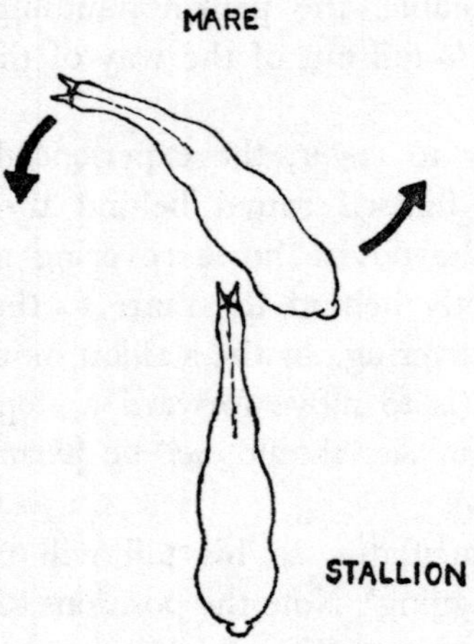

As the stallion dismounts turn the mare's head to the left
this causes her quarters to move to the right out of harms way.

Fig. 24. Turning the mare

After covering, the stallion's sheath should be washed off with a mild antiseptic or a specially prepared stallion wash, which is obtainable from most equine veterinary surgeons (plate 31).

Once every two weeks, it is a good idea to collect a sample of the stallion's semen into a sterile honey jar for bacteriological analysis. This is best done just as he pulls out of the mare, then a few drops will usually come away from the end of the penis. (This sample is unsuitable for quality analysis as it usually consists of about 90 per cent seminal fluid.)

When the stallion has been led away, the service boots can be unbuckled and the mare encouraged to walk out of them. Her tail bandage should also be removed. She can then be taken straight back to her paddock or loose-box. It is now considered quite unnecessary to walk a mare after service to prevent her from straining, or to tip a bucket of cold water over her back or put some nettles under her tail. If she does 'urinate' or strain and lose some semen, there will be plenty left to fertilise her egg.

When you are travelling your stallion to visit mares, you will need to telephone the people who own the mare to say at what time you are coming, and so make sure that they will have the mare caught up by the time you arrive. You will need to take the following equipment with you:

Stallion bridle
Bridle for the mare, in case the mare's owner does not have one
Service boots
Twitch
Tail bandage
Mild disinfectant
Stallion wash

When you arrive, leave your stallion in his horse-box and select a suitable place for trying the mare; a stout field gate is usually suitable providing it has not got anything solid immediately behind it which the stallion could catch should be kick out suddenly in excitement. Also, choose a suitable site for covering – this must be out of sight of any neighbours and have a non-slip surface.

If possible take someone with you to hold the mare, as covering is to a large extent a team effort and much safer for all concerned if the handlers know exactly what they are doing. If this is impossible, a little time should be spent explaining to the owner what to do with the mare during trying and covering.

When a field gate is used for trying, the mare should be kept head-on as far as possible and great care should be taken to ensure that she does not swing round, kick out and put a hind leg through the gate. A portable trying bar can be made (see Fig. 23) and taken, when visiting mares at home.

Any mares which are known to be difficult to cover should not be covered at home but the owners should be persuaded to bring them in to the stud, even if only for the short period when they are in-season. For this reason, it is almost essential that every stud should have its own horse-box or trailer, so that mares can be transported easily.

Really difficult mares are better tranquillized before covering lessening the risk of injury to stallion and staff. In these cases, it is often advisable to strap up the off fore leg, to prevent the mare from kicking. This is best done by slipping a leather strap round the mare's fetlock and passing the end of the strap over the withers and holding it on the near side. If a rope is used, it must be padded round the fetlock, as it would otherwise cut into the mare's leg.

Once the mare has stopped being in-season – 'gone off' – a card should be sent to her owner, giving the dates when she was covered:

Dear Sir/Madam,

Your mare .... visited

on ....

Yours faithfully,

10

THE FIRST SEASON STALLION

The first season stallion must be as fit for the covering season as the experienced horse. The exact system of management chosen must be dictated by the type of stud, breed of horse and the labour force available.

In the case of the thoroughbred horse out of training, he should be allowed a few months to let down before the season starts. His diet must be changed gradually from the high energy ration of the racehorse to maintenance only or one of about 14 per cent protein during the breeding season (see Chapter 14).

The amount of exercise must also be regulated to the requirements of his new life. In this respect, great care should be taken if the new stallion is to be turned out each day. Where the horse has been confined to a loose-box for any length of time, he must be galloped in a well-fenced paddock before he is turned out unattended. Many thoroughbred stallions spend their lives in close boarded paddocks but there is no real reason why stallions with suitable temperaments should not be turned out into well-fenced post and rail paddocks like any other horse.

Young horses which have been confined to a loose-box for any length of time, as is the case with a horse newly out of training or a show horse immediately after the season, will tend to gallop on being turned out for the first time. Where the animal is to be put into a post and rail paddock, there is always the danger that he will be unable to stop and either jump out of the paddock or crash through or into the fence – at the very least, bruising his shins. In order to prevent this from happening, enough people should be found to stand around the outside of the paddock fence and turn the horse as he gallops towards them.

The horse should be allowed to gallop so that he can work off his surplus energy while everyone is present but care should be taken to prevent 'tying up' when he is turned out for the first time. As soon as he settles, he can be left to graze quietly with only one person to watch him. If he has sweated up to any extent and it is not a warm day he should be brought in as soon as he has stopped galloping and before he gets cold. He can then be turned out each day, as a routine after a short period on the lunge. Once he has settled down, there will be no need to leave anyone to watch the horse.

The main object with a young stallion is to give him the maximum number of hand-picked mares which he can reasonably manage to get in-foal in his first season. The exact number will vary with the number of times he has to cover each mare, the age of the horse and time of the year when he retires to stud.

A mature horse six or seven years old should be able to cover twenty-five mares in his first full season whereas a four-year-old will probably only manage about fifteen. A few bad breeders will inevitably restrict the overall number of mares, as these will require more services per individual than those which are easy to get in-foal; therefore this type of mare is to be avoided when the total number covered is important. Veterinary examination before covering to check ovulation will also in theory reduce the number of services per mare.

The art of making a stallion mainly lies in getting the greatest number of live foals preferably in the country where he stands, for his first crop, which are of sufficient overall quality to enable the horse to have an above average number of winners in the show ring or on the racecourse, etc. Where the progeny are going to race, it is essential to have a large proportion of sprinter-bred mares which are likely to produce early maturing off-spring which in turn will win their races early in the season as two-year-olds.

Where there is an opportunity to get a very good mare to the horse in his first or second season it is often worthwhile offering a free service, as a mare which constantly produces good winners no matter which stallion she goes to, should be an enormous help to the horse. One above average winner, in his first crop together with some other average winners, should be sufficient to make the horse for a few seasons. The flow of winners must be maintained throughout the horse's career for him to be hailed as a successful stallion.

When you are introducing the young stallion to stud work, it is probably quite a good idea to let him tease a few mares – this should help to awaken his natural instincts and make him a little keener when it comes to covering his first mare. Even if it is intended that the horse will run with his mares during the season, at least one mare should be covered successfully in-hand before he is turned out. The first mare in all cases must be fully in-season and known to be very quiet. Ideally, she should be a little smaller than the horse and not stitched. A barren mare is ideal for this, as she tends to stand more quietly for longer periods than a foaling mare would, as after a short period she would start worrying about her foal, however quiet she is normally.

Some maiden horses take at least an hour or more to cover their first mare. Where the horse is very slow it is sometimes a good idea to take him back to his loose-box and leave him for an hour and then try again after which time he may be much keener.

For the first covering, allow yourself plenty of time and be patient. Patience is without doubt the most important factor with a young stallion. Get the mare ready for covering as usual, bring the horse up to the mare and let him tease her, the natural sex urge will cause him to draw but do not let him attempt to mount until he is fully drawn. When he is ready, position him exactly behind the mare. The greatest tendency with novice stallions is to mount from the side and get both front legs hooked over the mare's back or neck. This can be avoided mainly by continually positioning the stallion behind the mare and having two people on the mare's head so that they can push her round so that she is maintained in the correct position for covering every time the horse attempts to jump.

If the stallion makes a mess of mounting the mare, do not be tempted to pull him off backwards but rather walk the mare out from under him. In this way you will avoid any danger of pulling him over backwards, while he is in the rearing position. If space is limited – as in some covering yards – always walk the mare round and follow her at a safe distance with the stallion; this will avoid any danger of an excited young horse kicking out at the mare and damaging her or her attendants. Encourage the horse to mount by making him put his head on the mare's quarters.

When the horse jumps, it is often better not to help him to find the vulva as this usually only confuses him – that is, unless he is getting frustrated then some guidance will be necessary. Nature has equipped him with a very efficient system of finding the target. Some stallions find difficulty in dismounting from their mares for the first few times. Because of this, a soft landing must be ensured in case the stallion actually falls off his mare. In cases where the horse is reluctant to dismount, the mare should be walked forward and out from beneath the stallion.

After covering, wash the horse off in the usual way and if he has sweated up with excitement walk him round for about ten minutes to cool off before putting him back in his loose-box.

A horse which is going to be turned out with his mares for the season should be allowed to cover at least twice more, or until he is reasonably proficient at covering in-hand, before he is allowed to run with his mares. When turning a stallion out with mares for the first time, it is advisable to cover a quiet fully in-season mare in-hand and then turn her and the stallion out with the other mares immediately afterwards. He will then tend to concentrate on this mare and settle quicker than he would otherwise.

If at all possible, a novice horse should not be asked to cover more than one mare a day at least for the first month. Intially, it is not a bad idea to cover two mares a day for three or four days, followed by three or four days off, so that the horse will become keen and proficient at his new job fairly quickly.

Some studs do not accept any maiden mares to a first season stallion, as these mares are generally thought to be more difficult to cover. It is considered that a kick early in a stallion's career could make him a shy coverer for the rest of his life. However, maiden mares are usually much easier to get in-foal than some old barren mares and far less likely to be infected. Therefore, when your major concern is to get a large number of progeny in the first crop, it is often good policy to accept a few maiden mares.

These mares should not be covered until the horse has successfully covered several older quiet mares. Ideally, maidens should be 'bounced' by a teaser before they are given to the young horse. This is done quite simply by allowing an older horse or teaser to mount but not cover the mare, this gets the young mare used to the stallion's attentions and to being mounted before actual covering by the young horse takes place.

It is a good idea to accustom the young stallion to being handled between the hind legs and in the region of his sheath so that he will be easy to wash. It is essential that all stallions are kept clean and their sheaths are lubricated with liquid paraffin or vaseline to prevent them from becoming rough. The loose skin at the top of the sheath must also be washed, as lumps of grease and dirt will soon collect there if this is not attended to regularly. Stallions which are running out with their mares should be caught up and washed every week at least.

As an aid to ascertaining which mares are in-season and have been covered, a stallion which is turned out with his mares can have Sheep Raddle smeared on his chest, the colour being changed about every two weeks.

18. Felt covering boots are put on the mare's hind feet to lessen the chance of damage to the stallion should she kick

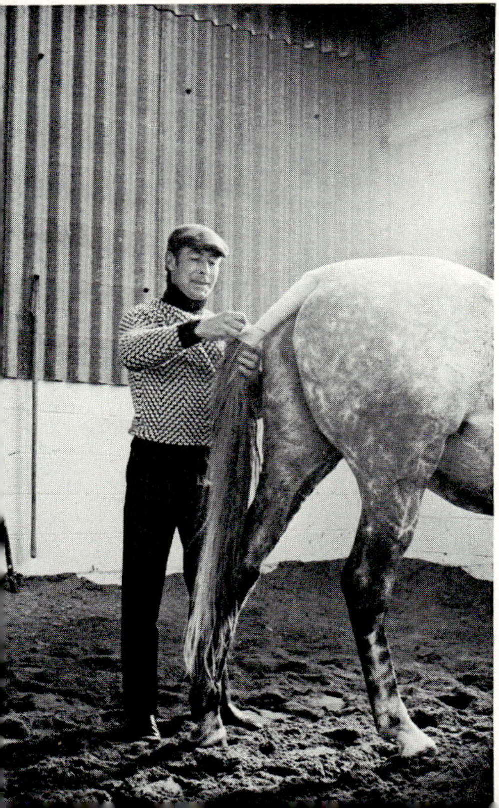

19. A tail bandage is put on the mare. Note the length of the twitches in the background – a short twitch is dangerous

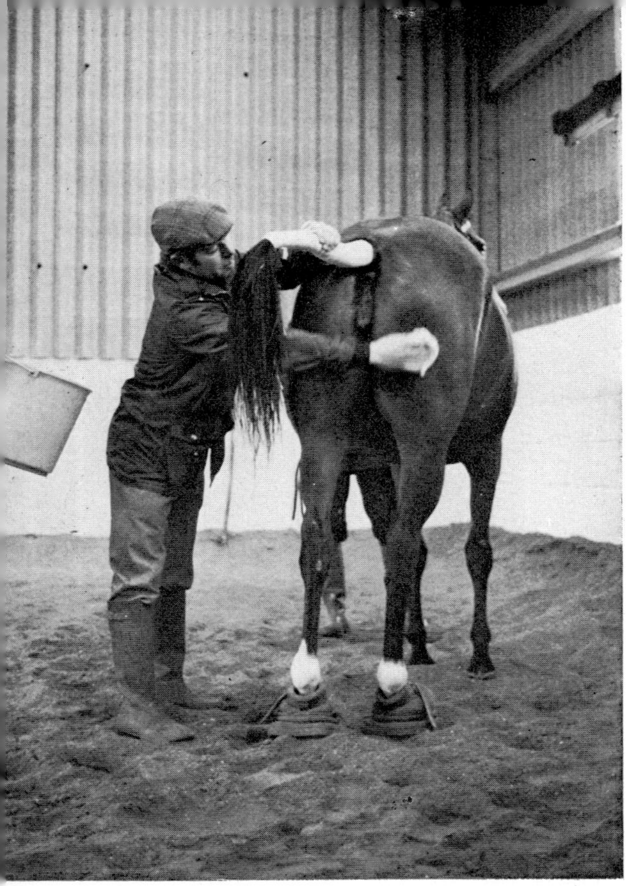

20. The mare's vulva and hindquarters are sponged down with a mild disinfectant solution

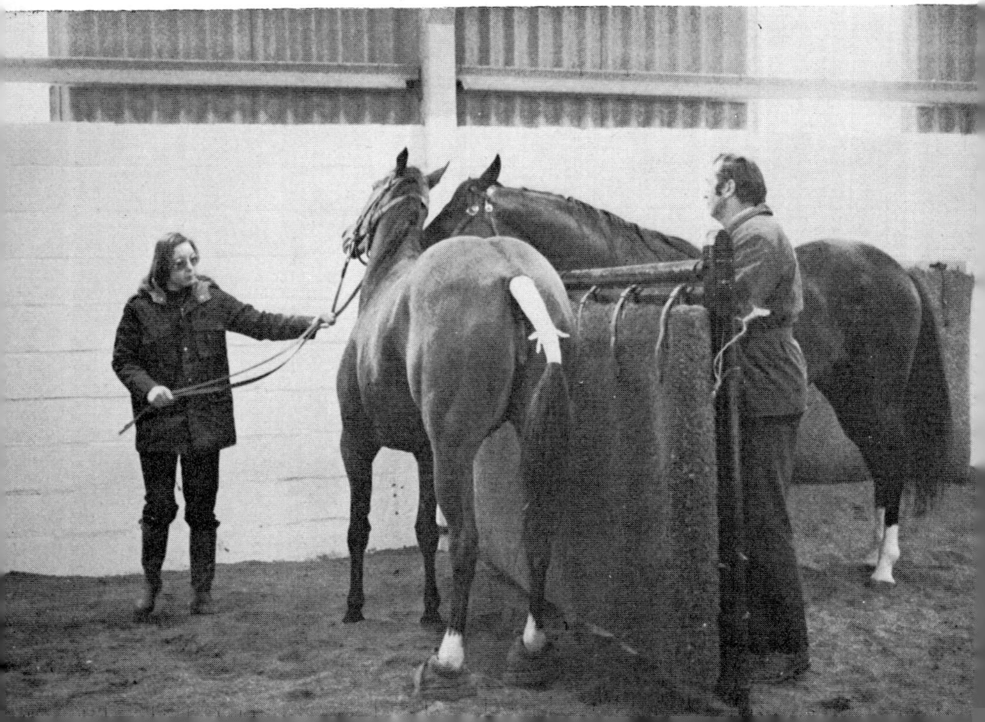

21. As a final precaution against accidents, the mare is re-tried immediately before covering. The trying bar, in this picture, is swung back against the wall before covering takes place. See diagram on page 76

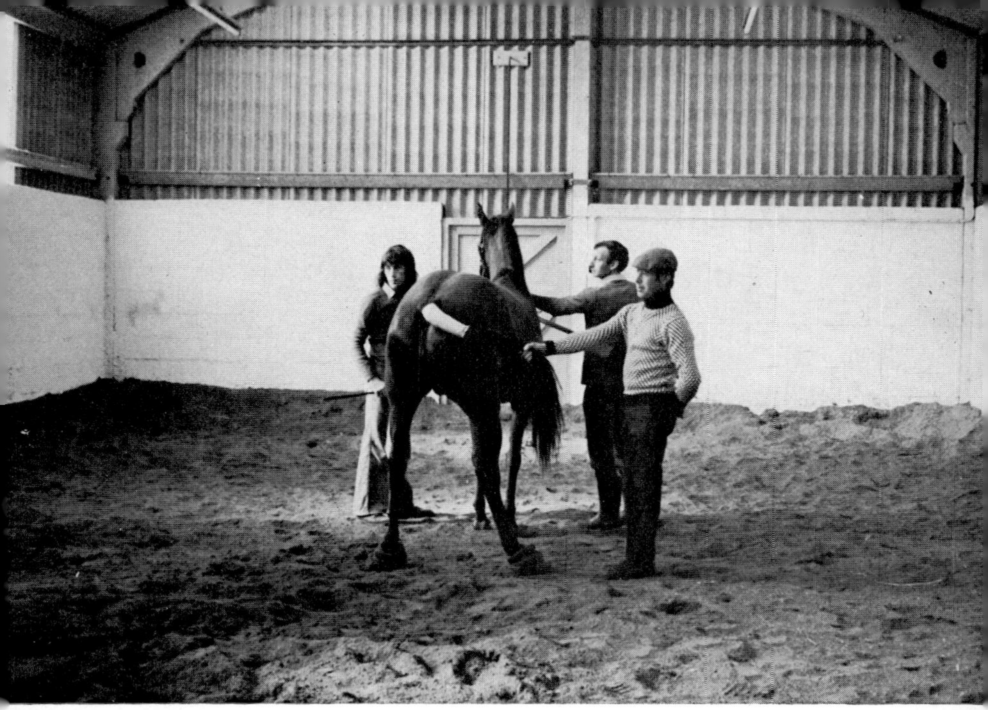

22. The mare is ready for covering. Her tail is held to the off-side out of the way of the stallion. 23. The stallion is brought up to the mare on her near side. The mare's head is kept slightly to the near side so that her hindquarters will move away from the stallion should she start to play up

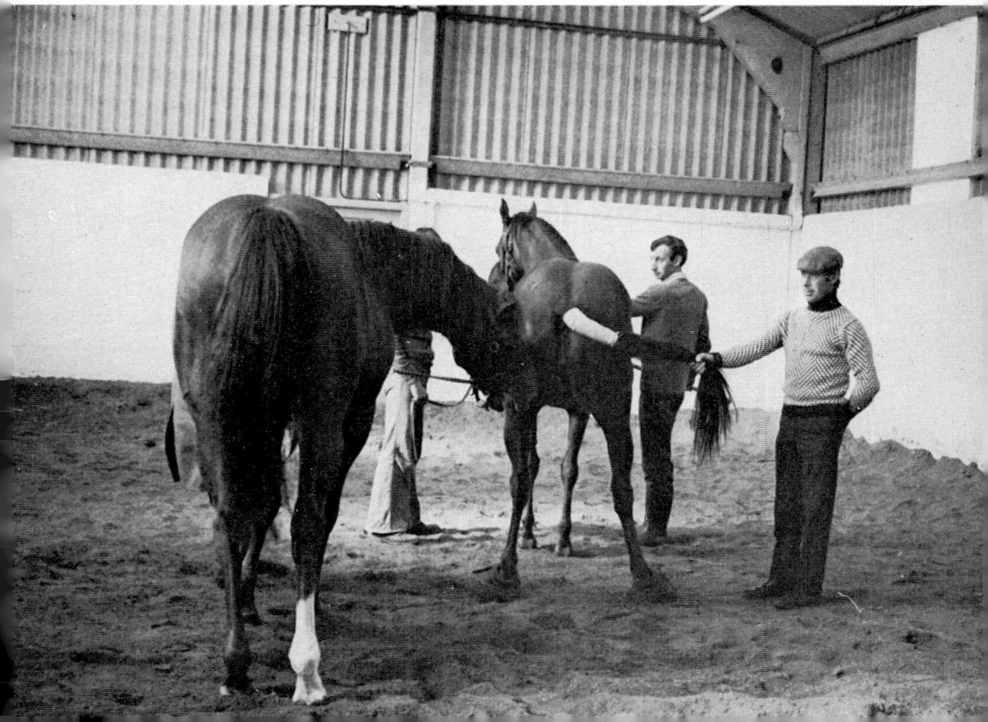

24. When he is fully drawn, the stallion is allowed to mount.
25. A hand placed on the stallion's front legs will often help to steady him

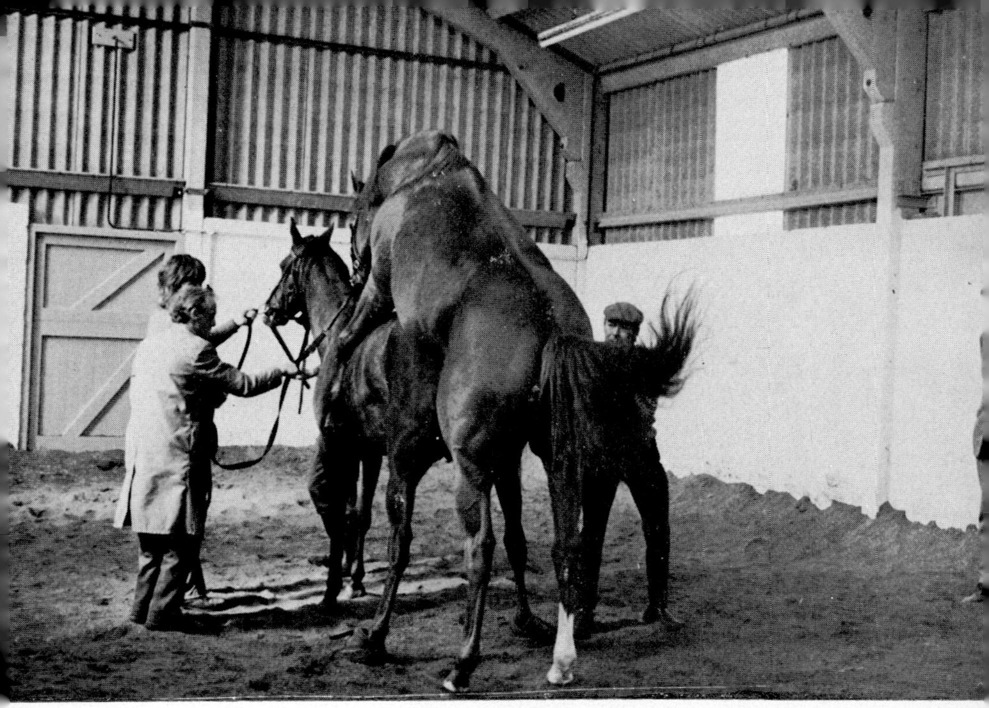

26. The mare should be encouraged to walk forward a step or two as he covers, or held quite still according to the stallion's preference. 27. When ejaculating, the stallion's tail will 'flag' (move up and down) which is a rough indication that he has covered his mare successfully

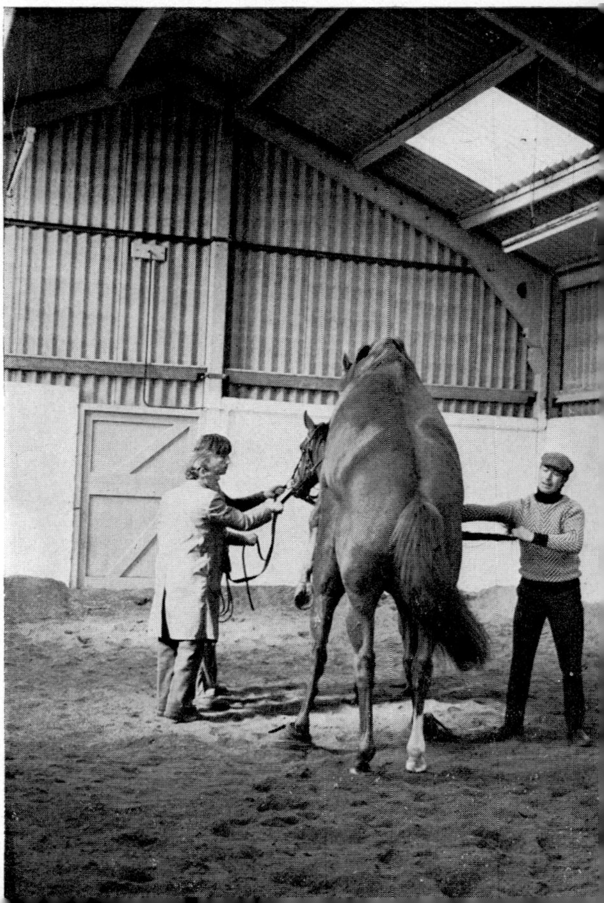

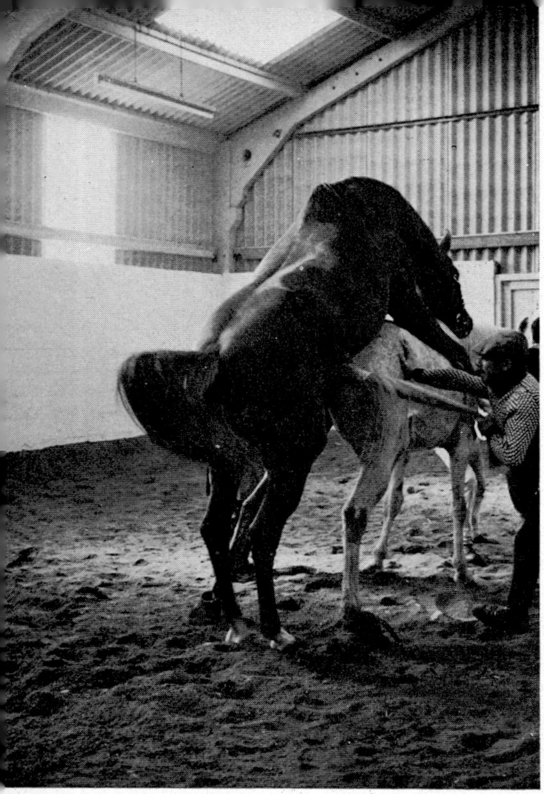

28. Care should be taken that the mare does not move sideways while the stallion is covering since this can cause damage to the horse

29. When the stallion has finished covering, he may rest on his mare for a few minutes or dismount immediately

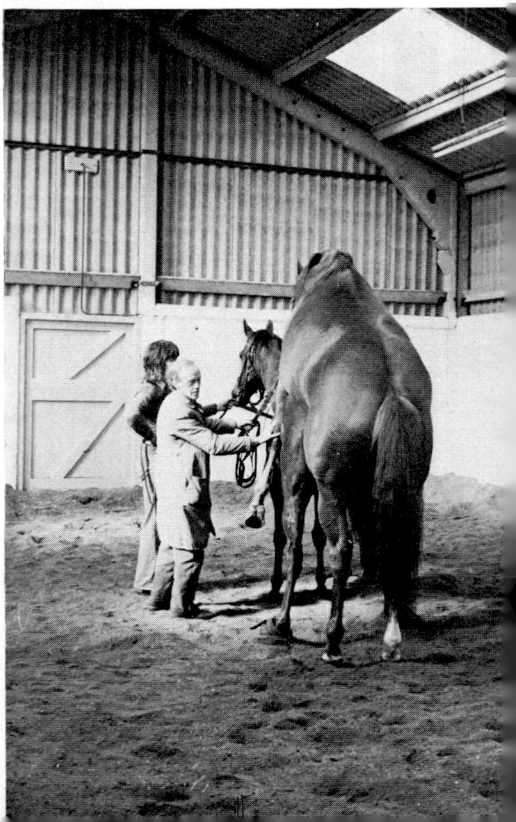

30. The moment the stallion starts to dismount, the mare's head is turned towards the stallion, moving her hindquarters to the right (see diagram on page 88). The stallion is backed away from her to prevent any chance of his kicking the mare or her attendant

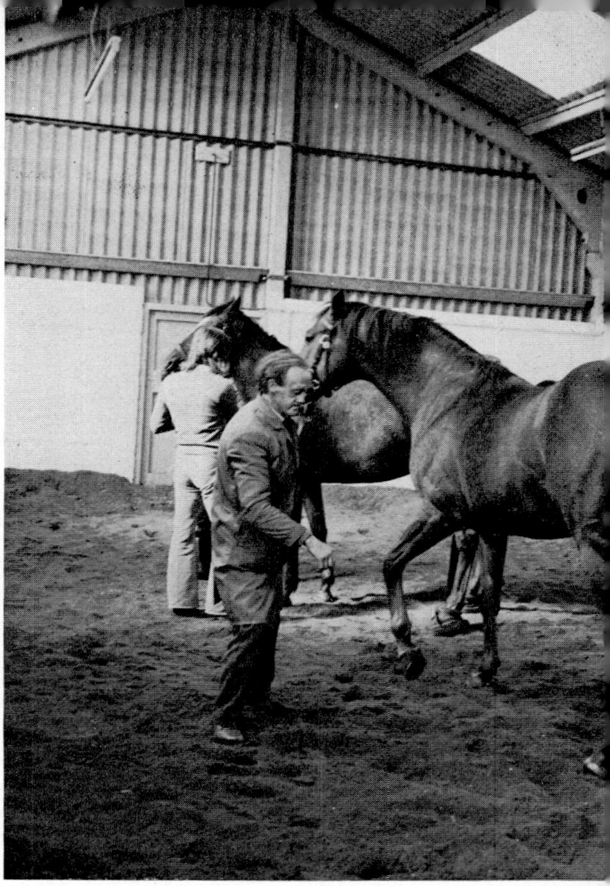

31. The stallion is washed down with a mild disinfectant solution.

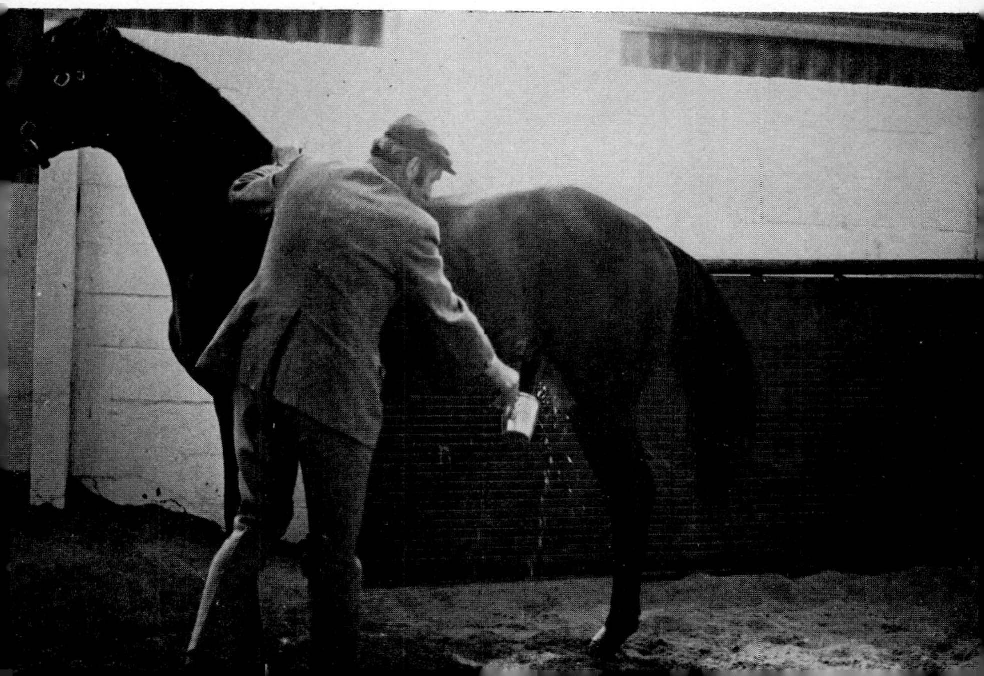

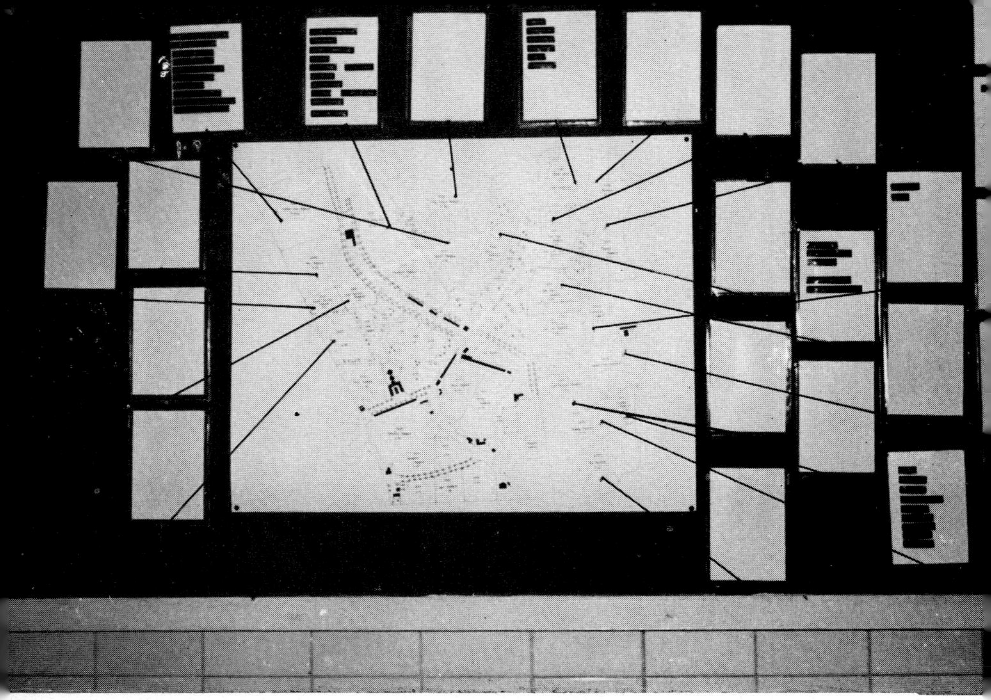

32. Paddock records: a coloured tape joins a paddock, shown on the map of the stud in the middle, to a plastic envelope at the edge which contains full details of that particular paddock. 33. An ideal stallion yard with a pleasant outlook

11

CONDITIONS ASSOCIATED WITH COVERING MARES

BACTERIAL INFECTIONS

The most common bacterial infections which are associated with covering mares and which can lead to abortions or the birth of sick foals are:

Streptococcus
Staphlococcus
E. Coli
Pseudomonas
Klebsiella

Streptococci are often responsible for abortions in early pregnancy. These organisms together, with Staphlococci, are normally present on the coat and skin of all horses and can therefore easily gain entry to the vagina during covering. However in the normal healthy mare, nature provides a satisfactory shield against infection, which is usually at its best during the period when a mare is in season. If there has been any damage to the vagina or cervix due to a difficult foaling, or if the mare has a sloping vulva or dropped cervix, her natural defences will be broken down or weakened, with the result that infections may be able to gain entry.

E. Coli on the other hand, are often responsible for abortions in the second half of pregnancy. Together with Pseudomonas, they are natural inhabitants of the gut and can therefore gain entry to the reproductive tract, either at the time of covering or in the case of mares with slack vulvas at any time. Mares with slack vulvas should be stitched: 'Caslicks's' operation.

Klebsiella is a specific venereal disease which is caused by a highly contagious bacterium. Infected mares and stallions usually show signs of a discharge. In some mares, the infection may persist despite antibiotic treatment.

Klebsiella, potentially can cause complete sterility in both the stallion and his mares. An infected stallion should not be allowed to cover any mares until he has been declared clean by a veterinary surgeon, which may take several weeks or even months, possibly resulting in the loss of a whole season. (See chapter 17 on insurance.)

Prevention is far better than cure, for this reason *all mares* should be swab tested before service by the stallion. *Not every infected mare shows signs of a discharge* therefore swab testing is very important when barren and barren-maiden mares are concerned (that is, any mare which has been covered before but which has not had a foal that year, but also including any mare which has foaled but has been previously covered by another stallion in the same season).

All stallions should have the penis and sheath swab-tested at the beginning of the season and ideally at two week intervals throughout the season, and/or a sample of their semen collected into a sterile jar for bacteriological examination every fortnight, to safeguard against outbreak of infection.

DAMAGE TO, OR RUPTURE OF, THE VAGINA OR CERVIX

Damage to the vagina or cervix can be caused when a large stallion covers a small mare. Where there is a great discrepancy between their sizes, the stallion's penis may be so long that it will tear the mare's cervix, which may effectively prevent her from breeding again.

This can be prevented to a large extent by using a 'breeding roll': this is a cylindrical padded bag with a stick running through the middle which acts as a handle. The roll should be covered with a polythene bag before use, which can be washed or destroyed after each service, to prevent the spread of infection. At the time of service, once the stallion has mounted the mare, the 'breeding roll' is pushed between the stallion and mare above his penis, and so prevents him from getting as close to his mare as he normally would, and thus reducing the risk of damage to the mare.

Another possible cause of a ruptured vagina, is when the stallion's sheath and mare's vagina are not on the same level; in this case, the penis will enter the mare at an angle and could therefore cause damage to, or a rupture of, the vagina wall. To avoid this a mound can be made for the smaller animal to stand on, see page 87.

If a mare bleeds profusely after covering, it is always better to ask your veterinary surgeon to examine her to make quite sure that no damage has been done. A tear in the vagina wall may need treatment. In the case of genuine maiden mares, bleeding from the vulva after their first service is quite common. This is only the result of the hymen being ruptured by the stallion during service. Some people like to break the hymen themselves, but this can be a very dangerous occupation. Although some normal mares have very tough hymens which need to be broken down manually, in very rare cases abnormal mares have no uterus and to break the hymen in these cases might lead to the death of the mare. So where you are dealing with visiting mares, it is far better to ask your veterinary surgeon to do the job for you.

The fact that a mare has haemorrhaged should not in itself have any effect on her chances of conception unless some severe mechanical damage has also occurred, such as a torn cervix.

FUNGAL INFECTIONS

Fungal spores are usually airborne and it is well-known that such things as dusty hay and straw contain millions of these spores. Therefore covering should not take place in a straw yard, nor anywhere where the atmosphere is very dusty. If the covering yard has a peat or sand floor, this should be kept damp throughout the covering season, in order to keep the dust down to a minimum.

Care should also be taken when mucking out mares, which are kept in for some reason. When shaking up the straw, one should keep the fork as low as possible, so that a minimum number of dust particles are shaken up into the atmosphere. This will help to keep down the incidence of broken-wind as well as abortions.

Uterine fungal infections nearly always result in abortion or the birth of a weak foal. Where a fungal infection is present, the afterbirth usually has a very characteristic creamy-coloured matt surface. Infection can occur at covering time, but in the case of a mare with a slack cervix, it may gain entry to the uterus at almost any time.

RECTAL DAMAGE

This accident can only happen when the stallion is not supervised and controlled during covering. It is more common in mares with sloping vulvas, particularly in older mares where the anus is very sunken.

Very occasionally the stallion may enter the rectum instead of the vagina, causing severe damage to the mare, which may result in rupture of the rectum and tearing of the anus which in turn may be followed by a fatal peritonitis.

If you call in your veterinary surgeon the moment the accident happens, he may be able to save the mare by means of surgery.

'SPOTS'

This is the colloquial term for infection with an equine herpesvirus, which is otherwise known as *Coital exanthema*.

The virus is from the same family as that which causes virus abortion, but there the similarity ends.

The organism is usually picked up from an infected mare by the stallion and passed on to clean mares at covering time. The stallion usually develops small spots on his penis which may become secondarily infected with bacteria and form small ulcers, which exude serum. The symptoms in the mare are similar: the spots becoming apparent on and around the vulva, when the spots heal they leave small hairless areas which lack pigmentation.

When an outbreak occurs, the stallion must be rested immediately and your veterinary surgeon called, as any mares covered at this time would probably become infected with the virus, which in turn would make them susceptible to secondary infections with such organisms as E. Coli, Staphlococci and Streptococci. The stallion will usually run a temperature at first and will often seem off colour for several days. The spots will usually disappear in about ten days.

12

SOME DISEASE CONDITIONS OF YOUNG FOALS

BARKING AND WANDERING

This is an obscure condition in new-born foals, in which the foal appears perfectly normal at birth but sometime afterwards, in some cases it will start to make a noise like a small dog barking and will show marked signs of respiratory distress. If the foal is already lying down he may jerk his head up and down and move his legs without being able to get up; if he is already on his feet he will probably show definite symptoms of blindness: he will wander around aimlessly, bumping into the walls and continually calling for his mother without apparently being able to recognise her.

The foal should be placed on a blanket under a lamp for warmth if possible and veterinary aid must be sought immediately before there is permanent brain damage.

The exact cause of this condition is obscure, it may be attributed to several factors such as infection by the germ *Actinobacillus equuili*, or other organisms such as those which can cause septicaemia. Lack of oxygen at the time of birth associated with premature breaking of the cord or fractured ribs sustained at the time of birth, could also possibly cause this condition.

This is a very rare condition except in Thoroughbreds.

ENTROPION

This is a condition in which the foal exhibits a chronic watering of one or both eyes. On careful examination it will be seen that the bottom eye lid has turned inwards and the lashes are irritating the eye.

This is a job for your veterinary surgeon and he may wish to insert some stitches into the bottom lid which will make the lid turn outwards in the normal manner, these are left in place for about two weeks.

HAEMOLYTIC DISEASE

This is a blood disease of the newly-born foal, in which the chief symptoms are anaemia and jaundice. The disease is caused by a blood group incompatibility between the foal and dam and is analagous to Rhesus babies as seen in human medicine.

The condition arises when a foal, while still in the womb, inherits a blood group from its sire which is incompatible with the dam's blood. The mare may then become sensitised to this blood group and produce antibodies against it. She concentrates these antibodies in her colostrum and on suckling they pass into the foal's blood stream; the antibodies then combine with and destroy the foal's red blood cells, which results in severe anaemia and jaundice.

If you own a mare which has had a foal with haemolytic disease or if you want to make sure that your mare does not have a foal with haemolytic disease, then about three to four weeks before she is due to foal, ask your veterinary surgeon to take a sample of her blood. He will send this away to a laboratory to be tested against samples taken during the succeeding two weeks, for an increase in antibody level. Any significant increase in level would denote that the foal will be at risk should it receive any colostrum from its own dam.

If it has been determined that the mare has an increased antibody level and that the foal when it arrives will therefore be at risk, the main preventive measures to take are:

(i) Obtain a foal size muzzle – this must be strapped on to the foal's head the moment he gets to his feet, to prevent him from drinking any of his mother's milk and must be left on for thirty-six hours. If all access to colostrum is withdrawn the disease cannot occur.

(ii) As the foal must have some colostrum an alternative supply should be provided, this can be obtained at any time from another unsensitised foaling mare and kept in a deep freeze until ready for use, it should be warmed to blood heat before feeding. Some veterinary surgeons in equine practise keep a colostrum bank for emergency use.

The mare must of course be milked out at least six times a day to prevent her from drying up before the foal is allowed to suck. This milk must be discarded.

The clinical signs of this condition are that the foal will be seen to yawn, appears sleepy and short of breath due to lack of oxygen in the blood, from the destruction of the red blood corpuscles. The mucous membranes are yellow. This condition usually becomes apparent within the first thirty-six hours after birth.

The foal unlike the human baby receives no immunity while in the womb, so if the foal does not receive any colostrum from its own dam *no* haemolytic diseases can occur. By examining in late pregnancy for antibody, a foal need never suffer from this disease.

MENINGITIS

This is the word used to describe inflammation of the membranes covering the surface of the brain, and/or spinal cord. It can be a sequel to severe infections of the upper respiratory tract, e.g. Strangles.

The affected animal usually becomes dazed and is liable to

walk into objects in its path. It appears ill – the temperature, respiration and pulse are higher than normal. Veterinary help should be called in immediately as this is a very serious disease and only very careful nursing can effect a cure.

NAVEL OR JOINT ILL

This is a disease characterised by swelling around the region of the navel and/or swellings in the leg joints accompanied by marked lameness in the affected limb or limbs.

The germs which cause this disease are thought to gain entry at the time of or soon after birth through the navel cord, or even from the mare before birth. There is particular danger when the cord is broken prematurely, as the umbilicus does not have time to close up immediately as it does when the cord breaks naturally. For this reason, as well as for reasons of loss of blood, the cord should be left to break naturally and maximum cleanliness should be observed at the time of foaling to cut down the possible risk of Joint Ill occurring, from this point of view it is far better to foal a mare outside in a field than in a dirty loose-box.

The symptoms of Joint Ill are usually that the foal goes off suck and appears dull and listless, the temperature rises to 103–105°F. and the foal will tend to spend most of its time lying stretched out on its side.

The navel may be wet and oozing a blood-stained fluid or it may be dry, swollen and obviously painful, due to the formation of abscesses.

Some of the foal's joints may also be swollen, tense and hot, if left untreated these will eventually burst and discharge blood-stained material, the foal will gradually get weaker and die, in some cases the abscesses are also found on the internal organs in which case death is very rapid.

Prevention mainly consists of maintaining scrupulously clean surroundings at foaling time and a liberal application of sulphonamide powder to the cord the moment it breaks.

Should you suspect that your foal might have Joint Ill, your veterinary surgeon should be called in immediately. No other mare should be allowed to foal down in the same loose-box as one that contained a foal with Joint Ill unless it has been thoroughly disinfected first, as this is a highly contagious disease – and very difficult to cure unless treated at once.

PERVIUS URACHUS

In this condition a continuous trickle of urine from the navel will be observed, which will probably take the hair off the lower regions of the hind legs.

Before birth the foal lies inside the sac known as the Amnion and is immersed in Amniotic fluid, this is surrounded by the chorio-allantois membrane which contains allantoic fluid, this is the foetal urine which has collected before birth, the urinary bladder opens directly into chorio-allantois which thus prevents over-distension of the bladder before birth, at the time of birth however, the umbilicus closes.

In the case of pervious urachus the urachus continues to have direct access via the navel cord with the outside and does not close up in the usual way.

This condition is sometimes self curative but if it does not seem to improve within two weeks your veterinary surgeon should be consulted as treatment may be necessary. Vaseline should be applied to the hind legs to prevent blisters forming from the action of the urine on the skin.

PNEUMONIA

This is inflammation of the lungs with a resulting breakdown of the lung tissue which greatly impairs respiration. A variety of germs: bacteria and viruses, are able to cause pneumonia in susceptible young foals.

Predisposing causes of pneumonia may be exposure to cold, wet weather, or from stuffy and badly ventilated stables.

It can also be produced by careless drenching – allowing some of the medicant fluid to go down on to the lungs.

The possible symptoms of this disease are: a high temperature, fast laboured respiration, the foal will go off suck and there will probably be fits of shivering, there may be a nasal discharge. In the case of pneumonia the care taken in nursing and management is of utmost importance as a relapse can take place at any time if the standard of care is relaxed.

A veterinary surgeon must be called in at the very beginning if a good chance of recovery is to be expected.

RETAINED MECONIUM COLIC

Meconium is the brown or black hard excretory material which collects in the bowels before birth. It should be passed within a few hours after birth. The passing of the meconium is accelerated by the purgative action of the colostrum. The meconium is thought to consist of bile and debris from the intestines which has collected before birth.

It is very important that the meconium should be passed within a few hours of birth to allow normal digestion to take place. If the meconium is retained, the foal will probably carry its tail higher than normal, it may be seen standing in a crouched position with its back arched straining; in a more advanced stage it will show very definite signs of colic: i.e. kicking up at its stomach, rolling and even lying on its back with its legs in the air in an effort to relieve the pain.

Retained meconium and constipation in the foal is often associated with a costive diet in the mare, therefore as far as possible the mare should receive some green stuff, carrots or mashes every day as well as adequate exercise.

In the early stages of retained meconium colic the foal can be given a dose of liquid paraffin. It is probably a better plan to call in your veterinary surgeon immediately you see the first symptoms.

An enema of warm soapy water does sometimes help to relieve this condition.

SCOURING

This is another word for the condition known as diarrhoea, a symptom of several diseases or errors in feeding.

The foal will often scour when the mare comes in-season due to some change in the composition of her milk at this time, which has an irritating effect on the bowel, this usually stops once the mare goes out of season. In this case the foal does not go off suck. A useful tip is smear vaseline round the foal's hindquarters to prevent the hair from coming out, this is especially important if you intend to show the foal.

Scouring at other times in young foals should always be treated immediately by a veterinary surgeon, as young foals have very little reserves and are soon pulled down in condition. Some cases of scouring are due to bacteria which can be identified by taking a swab from the rectum and growing the organisms in the laboratory. Any foal which is seen to be scouring should be isolated immediately as this disease can spread to other foals very rapidly.

Draughts and sudden changes in temperature can be predisposing causes of scouring. Other types of scouring can be due to digestive upsets.

Of major importance in all cases of scouring is an adequate intake of fluid to replace the loss to the bowel, at the same time certain vital substances known as electrolytes which are lost during scouring must be replaced. This is often done by the veterinary surgeon by means of a stomach tube. If the fluids are not replaced in adequate amounts the foal will die as a direct result of fluid loss. The foal's tongue usually becomes coated and the eyes may sink into their sockets, there is a marked loss in condition and wasting of the muscles.

Treatment usually consists of the administration of antibiotics by injection or by mouth, and foals usually recover without any permanent ill effects. Injection of some of the mare's blood into the foal will sometimes give an added boost to recovery due to the antibodies this blood will contain.

WEAK AND DEFORMED LEGS

Foals like babies can be born with various weaknesses and deformities of their limbs.

Some of those more commonly seen are :

(a) Over long weak pasterns, where the foal walks with his fetlocks touching the ground. This condition is usually self curative, as the foal becomes older and stronger the pasterns tend to strengthen. Reducing the length of the toes of the affected legs does seem to help.

(b) In the case of contracted tendons the foal may only be able to walk on the tips of his toes, in very bad cases the foal may be entirely unable to get up, if the contraction is very bad the foal may have to be put down, but in some cases an operation to cut the tendons and straighten the leg, may be successful, in other cases splinting the leg will often work well.

(c) When foals are born with twisted legs which do not start to straighten within a week of birth, or where their legs start to twist soon after birth, your veterinary surgeon should be called immediately as the longer these weaknesses are left the worse the condition will become.

Feet which turn in or out can be corrected by a good blacksmith to a large extent but where the whole leg turns a splint will probably be needed to correct the fault.

13

CONDITIONS ASSOCIATED WITH FOALING AND LACTATION IN BROOD MARES

BRUISING AND LACERATION OF THE VAGINA AND VULVA

This is a fairly common complication during foaling, and can occur during the birth of a large or abnormally presented foal, especially in very young mares or mares with a small pelvis. Laceration and tearing of the vulva will occur if a stitched mare is not cut enough to allow the foal to be born.

Damage by the foal as it leaves the birth canal will soon be noticeable: the lips of the vulva swell, appear congested and painful. The vaginal wall will also be reddened and somewhat inflamed. Examination of the mare in this case after foaling by your veterinary surgeon, is a wise precaution, to help prevent further trouble later on.

The risk of infection to the damaged tissue will be increased and this could lead to a metritis and cervicitis, possibly resulting in an infertile mare.

Mares which have sustained some bruising and lacerations at foaling should not be covered at the foaling heat, but should be left to some future date when they have healed up sufficiently. These mares often show signs of an acute and sometimes profuse vaginal discharge, with veterinary treatment this usually clears up quite soon and the mare will probably, then be ready to serve next time she comes in-season.

EQUINE VIRUS ABORTION

This disease is endemic in America but sporadic outbreaks do occur in the British Isles almost every year. Vaccination of all foaling mares is routine practice on most well run studs in the States, but as a live vaccine is used for this purpose, vaccination is not permitted in the British Isles.

This virus is identical with that causing 'snotty noses' in young horses – a condition usually seen during the autumn and winter. Abortions occur in infected mares between the fifth and eleventh months of gestation, but most occurring during the ninth and tenth months. Affected foals carried to full term are usually born dying or in a very weak state. The incubation period appears to be between twenty and ninety days.

Immunity to respiratory infection caused by this organism only appears to last for three to five months, but immunity to abortion appears to last longer, as mares will often show evidence of respiratory disease when in contact with infected youngstock but will seldom abort as long as their level of natural immunity is maintained by regular exposure to infection. Abortion only occurs in stock whose level of immunity has been permitted to wane, thus allowing the virus access to the foetus.

Abortion does not appear to occur in mares infected with the virus in the early periods of gestation. Where abortion does occur the mare does not make the usual preparations for foaling and seldom has any milk; she usually aborts quickly and easily. The afterbirth is seldom retained. The foetus usually appears very fresh and the bag is seldom broken.

The virus usually dies in about two weeks but on horse

hair it can survive for as long as forty-nine days. All bedding must be burnt and the loose-box thoroughly disinfected. The mare herself should be washed down with disinfectant; special attention being paid to her hindquarters and tail.

The foaling heat should be missed but infected mares can be covered safely at the next heat, when they will often conceive and foal down normally.

Your veterinary surgeon should be notified immediately if:

(i) A mare aborts;

(ii) A foal shows symptoms of pneumonia or dies within the first seven days of birth;

(iii) Any adult horse show signs of incoordination or paralysis or becomes unnaturally nervous.

To prevent outbreaks all animals from sales and from abroad should be isolated for a minimum period of one month.

When a case is confirmed no horse should be moved from the stud for a month after the last outbreak. After removal in private transport, the animal should be kept in isolation for a further period of two months, or ideally until foaling the following year.

HAEMORRHAGE OF THE UTERINE ARTERIES:

This condition usually occurs within a few hours of foaling. In the majority of cases there is no sign of any blood escaping from the vulva. When the mare is suffering from an internal haemorrhage and is losing a large quantity of blood, the mucous membranes (i.e. the gums and eyes) become pale and anaemic looking. The mare may sweat profusely and show signs of colicky pain. As she gets weaker, she will be unable to stand and may finally become unconscious.

In some cases the bleeding will stop automatically but in more severe cases of haemorrhage the outlook is very grave, however blood transfusions can sometimes be given, but in the case of severe haemorrhage these may probably be of no use.

This condition is more common in older mares, i.e. ten years and over which have had several foals and more so those with very dipped backs and dropped bellies, and is probably due to the stretching of the uterine ligaments and arteries. The actual haemorrhage being caused by rupture of one of the arteries to the uterus due to damage in the wall of the vessel.

METRITIS :

This is inflammation of the uterus which may be either acute or chronic. In the chronic form the mare becomes permanently infertile, while in the more acute form the mare may recover with time.

When acute metritis occurs immediately after foaling it is usually due to the introduction of septic bacteria (germs), into the uterus during foaling, these organisms are usually conveyed by the unwashed hands and arms of the attendants or from unsterile ropes and instruments used to assist the foaling. It can also arise from a portion of retained afterbirth, this is usually a piece of the non-pregnant horn (see plate 10). which remains attached inside the mare. Occasionally, due to the great weight of the portion lying outside the mare, the non-pregnant horn is torn through and is thus retained.

Metritis is a very severe condition in the mare and if left untreated is fatal. Your veterinary surgeon should be called in immediately the symptoms are observed. The mare becomes very ill and loses all interest in her foal, usually within the first two days after foaling. There is usually a blood-stained grey discharge from the vulva, which will soil her tail. In very bad cases the mare may appear tucked up and stand around with her back arched, in obvious pain.

As far as metritis is concerned, prevention is better than cure, extreme cleanliness should be observed at foaling time. If it becomes necessary to insert your hands and arms into the vagina during foaling you must make sure that they are scrubbed clean in a disinfectant and your nails are short and free from dirt. All instruments and ropes used for foaling must be sterilised, preferably by boiling for ten minutes, after which they should be left covered in the container in which they were boiled until ready for use. Loose-boxes used for foaling must be swept clean of dust and cobwebs and the straw used for foaling must be the cleanest available.

RECTO-VAGINAL FISTULA:

This condition is fortunately uncommon but may occur when due to abnormal presentation of the foal in the birth canal, one foot is pushed through the wall of the vagina and into the rectum. It is most common in first foaling mares.

The symptoms in individual mares will depend on the extent and position of the injury. In some cases the foot may be seen protruding through the rectum.

Where possible, the foal should be pushed back inside the mare. In any case your veterinary surgeon should be called in immediately and the mare got on her feet and kept walking round until he arrives, to prevent her from straining.

In most cases an operation to repair the tear can be performed with success.

RETENTION OF THE AFTERBIRTH:

In this case the afterbirth is not expelled in the normal way by the secondary birth pains, after foaling, but remains attached inside the mare. It must be removed within twelve hours of foaling or infection may set in, leading to a metritis and a possible general septicaemia with fatal results. In this respect the mare is far more susceptible than the cow, which can often be left for days with any apparent ill effects.

On no account should the attendant attempt to remove the afterbirth himself, as traction on the visible part of the membranes would nearly certainly lead to some of the attached portions, tearing inside the mare. These portions would be retained and decompose which would lead to an acute infection and probable death of the mare.

As a general rule of thumb, if, when the mare has foaled during the night, the membranes are still retained first thing in the morning, your veterinary surgeon should be notified immediately, before he goes out on his rounds. When he arrives he will need a clean bucket of warm water, some soap and a clean towel.

UTERINE PROLAPSE:

This can occur anytime up to about twenty-four hours after foaling. In some cases only the vagina appears outside the walls of the vulva but in other cases the whole of the uterus invaginates and appears hanging downwards, as a large fleshy mass, often reaching as far as the hocks. This condition can follow an easy foaling and is probably more common in mares with high set tails and slack ligaments round the vulva.

A partial prolapse of the uterus can occur internally when one of the horns turns in on itself, in this case the mare will show very definite signs of colic, she will sweat profusely and get down and roll in an effort to ease the pain. There will be no other visible sign that there is anything wrong. Your veterinary surgeon should, however, be called in immediately.

If the whole of the uterus is prolapsed, it must be kept clean and warm until the veterinary surgeon arrives, and great care should be taken to prevent its getting damaged in any way. The easiest way to do this is to place the mass on a clean sheet and remove all the particles of straw which may be adhering to it. Clean warm water may be sprinkled over the uterus to prevent it from drying up while you await the arrival of the veterinary surgeon, as the longer it remains outside the body the more swollen and dry it becomes.

Mares have recovered after a uterine prolapse but generally speaking unless treatment is prompt the outlook is grave.

14

FEEDING HORSES AT STUD

It has long been recognised that some stockmen, working, under identical conditions to others', will obtain infinitely better results using identical basic rations. These people are what is often referred to as 'born feeders'. Their secret probably lies in attention to detail, with special emphasis on the relationship of feed to condition, and a willingness and ability to look at each animal constructively every day and alter its diet as necessary. These people take a great pride in their animals, which always look just that little bit better than those of any other stockman. If you have such a person in your yard, you will not need to worry too much about any elaborate feeding programmes.

For the rest of us, a knowledge of the science of feeding is essential so that any ration can be brought within certain limits which have been carefully worked out to give optimum results under average conditions. To this end, many studs have changed over to cube feeding. Most firms, these days, market a wide range of balanced compounds for every stage and condition in a horse's life, from the foal, through weaning, to its emergence as a racehorse, hunter, etc. All the feeder needs to do is to obtain the necessary feeding instructions from the manufacturer and follow them to the last letter.

However, where better than average condition is required –

for example, when animals are to be prepared for the sales or show ring – an optimum ration should be utilized, if the animals in question are to look just the much better than their neighbours. So a working knowledge of both the science and practice of feeding is essential.

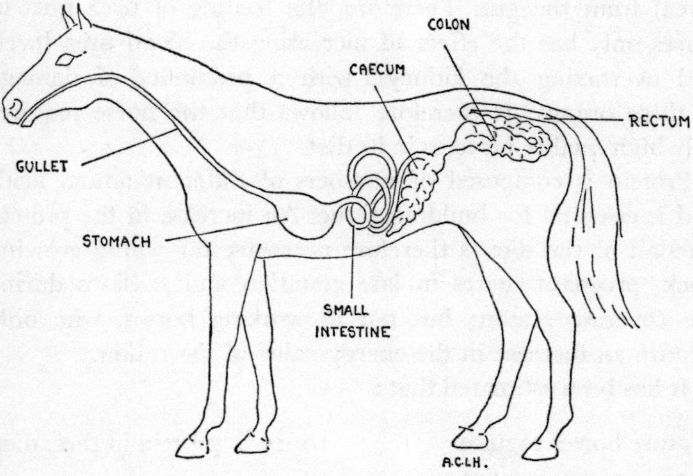

Fig. 25. The digestive system

A basic understanding of the digestive tract in the horse is essential to good feeding. The rate at which any feed travels along the digestive tract governs its digestibility and degree of assimilation. It has been estimated that, on average, it takes between three and four days for any feed to clear the digestive tract completely; but most food will pass through the stomach and small intestine in about twelve hours. The latter is the area in which protein, carbohydrates, fats and minerals are absorbed.

It follows, therefore, that the larger the quantity of food passing through these areas at any one time, the lower the degree of actual assimilation; therefore, it is better to feed horses 'little and often'. The large intestine contains bacteria which break down the fibre content of the ration to volatile fatty acids which are then absorbed. Any protein which has not been broken down into its amino acids and absorbed from the small intestine enters the large intestine and on reaching the blood stream will push up the ammonia and blood urea levels. It is now thought that the horse, unlike the ruminant, is incapable of utilizing non-protein nitrogen (urea) from the gut. Therefore, the feeding of urea nuts to horses only has the effect of increasing the blood urea levels and overtaxing the kidneys, with a possibility of damage to these organs. It therefore follows that the horse requires only high quality proteins in its diet.

Protein is composed of numbers of different amino acids and is essential for building tissue. An increase in the protein fraction of the diet is therefore necessary for young growing stock, pregnant mares in late gestation and stallions during the covering season; but not in working horses, who only require an increase in the energy value of the ration.

It has been estimated that:

Mature horses require	10–12% protein in their diet
Foaling mares require	
First two-thirds of gestation	12% protein in their diet
Last one-third of gestation	14% protein in their diet
During lactation	16% protein in their diet
Stallions during the covering season	14% protein in their diet
Yearlings require	20%+ protein in their diet
Foals require	14% protein in their diet

The protein content of the rations must be of the best quality and should be drawn from a wide variety of ingredients, to give the greatest possible range of amino acids in the diet.

In the author's opinion, therefore, a diet composed mainly of oats and hay does not fulfil the requirements of a well-balanced ration. A greater variety of feeding stuffs must be included. When selecting foods for a horse ration, these must be of the highest quality obtainable and must be presented to the horse in their most easily digestible form. These days with ever increasing costs of foodstuffs, the prices of the individual constituents of a ration must, also, be carefully watched. In many cases ingredients can be bought direct from the farm at considerably reduced rates, thus providing an overall reduction in feed costs.

Whole barley can be boiled up with linseed, without any need for rolling the grain. Whole field beans (boiled) can also be added to the linseed and barley for feeding to animals with a high protein requirement. Similarly, raw eggs can also be utilized where cracked or blood eggs can be obtained cheaply. Raw eggs, if fed, can be added to the milk before it is poured over the ration. Milk powder, although expensive, is also a very valuable and easily digested form of protein and is an almost essential addition to the diet of weaned foals. For ease of mixing it should be made up as per the instructions on the bag and used, to dampen the ration, in its re-constituted form. Flaked maize (which is high in starch) can also be added to the rations of young, growing stock, as it is very easily digested and helps to keep a ration open and palatable.

When these foodstuffs are incorporated into the ration, it is unnecessary, and probably a waste of money, to include any horse-nuts as you will only be duplicating the ration you are already feeding.

A foal will not consume very much solid food for the first few weeks of its life, but will rely on its mother's milk. During this period of maximum growth it has been estimated that the foal requires just over $\frac{1}{2}$ lb. of digestible protein for every 1 lb. gain in body weight, supplied initially by the milk alone.

The young foal has not developed a natural flora in its large intestine by the time it starts to eat, and therefore it should only have access to very high quality foodstuffs containing only easily digestible proteins. This should eliminate any carry-over of undigested protein to the caecum – thought to be a possible cause of early scours.

A ration must contain fats and carbohydrates as well as proteins. These are required to provide energy as well as maintenance. The carbohydrates are most important for producing energy; lack of which can produce stunted growth, loss of weight and condition. They are provided by the grain portion of the horse's diet and are used to produce energy; any excess is stored as fat or excreted. To increase the utilization of cereals, particularly in the case of older horses, the grain must be bruised, rolled, or fed cooked. Pregnancy does not apparently increase the energy requirements but lactation is thought to; as does covering, in the stallion. Therefore, the starch fraction of a stallion's ration should be increased immediately before and during the breeding season to a large extent, this can be supplied, from the hay, Unlike the horse in-training, this (and the reduced exercise) will help the horse to let down.

Fibre, although of less importance to the horse than to the ruminant, is nevertheless necessary. Young horses and those doing fast work should, however, have rations in which the carbohydrate fraction is low in fibre. When stabled, they should be fed good quality first crop hay ideally cut as young as possible.

Minerals are necessary for all classes of horses, and a reputable supplement should be added to the daily ration and salt licks put in every loose-box. Phosphorous should, however, never be fed in a higher ratio than calcium. Both these minerals are of major importance, particularly in the case of in-foal mares and youngstock. The minimum recommended ratio is 1.1 : 1.0. Most plants contain large quantities of phosphorous, but not all of this is in a readily available form. According to some authorities, both iodine and iron are sometimes found to be deficient in pregnant mares. It can be wasteful to feed calcium and phosphorus supplements without first checking to see if there is in fact any deficiency. This is best done by asking your vet to take random samples of blood for laboratory analysis. Some firms of proprietary horse feeds will have these tested for you free of charge if you use their food.

The assimilation of some minerals is governed by certain vitamins. The best example is Vitamin D (the sunlight vitamin), which is necessary for the proper utilization of calcium and phosphorous in the diet and, therfore, the production of sound bone. A deficiency of this vitamin will lead to rickets. Vitamin A is naturally available from the carotene in the grass and hay; a deficiency of this vitamin could only occur among horses stabled for any length of time without access to green herbage. Overfeeding vitamin A supplements can be harmful. Vitamin E (the fertility vitamin) is probably only of value to breeding and growing stock; it can be obtained easily in powder form for inclusion in feeds, but is very expensive. Vitamins can be administered by injection or by feeding, the latter being the less expensive method of the two, and the less likely to lead to over-dosing.

The practical aspects of feeding are based on the theories already described, which can then be adapted to fall in with any particular stud's availability of produce and routine of feeding.

As the visiting mares start to arrive in the spring, they should be segregated into their respective categories – namely, foaling mares, barren mares and maiden mares.

The foaling mares must be kept in at night and turned out during the day for exercise. Adequate exercise is of paramount importance where foaling mares are concerned, as lack of exercise can lead to complications at foaling. It has been demonstrated experimentally that improved feeding thirty-three days before foaling will increase the resulting milk supply but not affect the size of foal at birth. Mares in the last few weeks of gestation tend to eat considerably less, yet their requirements are steadily increasing with the size of foal. Therefore these mares must have smaller feeds of a higher nutritional quality than the other mares on the stud. These can take the form of regular linseed mashes and a high protein cube added to the usual oat ration. It is also a good idea to feed these mares three times a day, dividing their feed into three small parts. The first, given in the morning, at least one hour before turning out. It is most important to feed horses in the morning well before their normal time for going out, so that they will settle and eat their food. The second feed should be given on coming in and the third, last thing at night.

The barren mares can be turned out to grass day and night, the moment the weather improves and there is some spring grass; this will save loose-box room and labour. These mares should be given supplementary feed, until there is sufficient grass available for their maintainance. The easiest way to feed them is by using a Land-Rover to cart bales of hay and bags of nuts out to the paddocks. Although some studs use a tractor and trailer, this is possibly more dangerous as there is always the possibility of a mare being pushed over the trailer bar by the other mares. Allow one bale of hay for every two to four mares when necessary: $\frac{1}{2}$ cwt bag of nuts between eight mares according to the value of the grass. Drop off the hay and nuts in piles, across the paddock, allowing at least three more piles of each, than mares in the paddock, to prevent fighting. Choose the cleanest and most sheltered areas. You will soon be able to judge when to stop feeding these mares as they will stop clearing up the hay.

The maiden mares should be allowed out for exercise every day, but kept in at nights until the weather is warmer and there is plenty of spring grass about; then they can be turned out day and night. Their first night out should be warm and dry.

When feeding mares with foals at foot, care should be taken to ensure that the food is scattered across the full length of the manger, so that the foal can have a fair chance of getting some. When dealing with greedy mares, the foal should be provided with an alternative manger and, in very bad cases, the mare should be tied up while she is eating, to give the foal a chance; but care should be taken that she is un-tied afterwards.

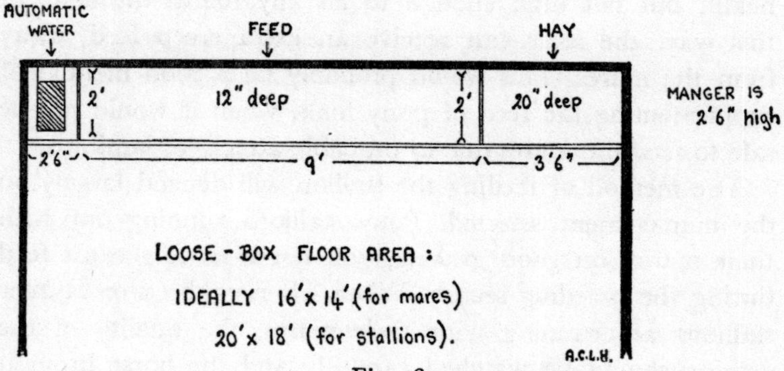

Fig. 26.

An ideal layout for stud loose-boxes, suitable for Thoroughbred, Hunter and Arabian type mares. Also eminently suitable for stallions, when the size of the loose box should be increased to 20 ft. x 18 ft.

Where accommodation and labour are in short supply, mares with foals at foot can be run out at grass day and night once the nights start to get warmer. A sheltered paddock should be chosen for these mares. When you are dealing with Thoroughbred, Hunter or Arabian type mares, these must be fed. They can be provided with wooden field troughs – one per mare, so that they can be fed each and every day.

These troughs may be made of wood measuring about 10 feet long and constructed on four legs: they should be approximately 2 ft. 6 in. high and the edges covered with metal strips to prevent damage by chewing. This is a popular method of feeding mares in the U.S.A. and Australia, which the author has used, with success, in England. The troughs must be located several yards apart to prevent fighting. In most cases, it will be found that the foals will stand on the opposite side of the trough to their mothers, thus avoiding being chased off while eating.

Another method, which has been used with success, is to construct a square fence round some field troughs. The fence is made just high enough to allow the foals to walk underneath, but not high enough to let any mares through. In this way, the foals can receive an extra creep feed, away from the mares. This would probably be a good method of supplementing the feed of pony foals, when it would not be safe to feed the mares due to probable attacks of laminitis.

The method of feeding the stallion will depend largely on the management selected. Pony stallions running out with their mares, on good grazing, will not need any extra feed during the breeding season. When Thoroughbred or Hunter stallions are running with their mares, the quality of the grazing should be watched carefully and the horse brought in for one feed per day when necessary. Always err on the side of feeding the horse, rather than waiting for him to lose condition before you start.

Stallions which are housed during the breeding season should receive a feed, at least one hour before they are exercised or given their first mare. Horses which are turned out during the day should be given three feeds, if they are covering two or more mares a day regularly. They should be brought in at lunch time and given a small feed and then turned away again, until they are required for covering or to be brought in for the night, when they will receive their last feed. Stallions kept in day and night, and only exercised on the lunge and led, should also receive three feeds per day, but should not be fed sooner than one hour before work or covering.

The stallion is the main reason for the existence of a public stud. Only the best quality feed should be given to the horse; the best hay and oats being selected and saved for his use Stallions confined to their boxes during the breeding season should be taken out to graze on the end of a lunge-rein, if time permits, and fresh grass should be cut for them every day. The grass must be cut and feed within an hour or so, otherwise it will soon start to heat up and ferment; if fed in this state it could cause colic. Whole carrots can also be fed.

However, care should be taken not to overfeed stallions otherwise they may develop laminitis. This can be of particular importance in the case of some small ponies.

15

NOTES ON STUD PADDOCKS AND THEIR MANAGEMENT

Grass is a food rich in vitamins and minerals. One of the major difficulties in its management arises from the fact that one has to deal with a mixed crop of grasses, clovers and herbs. It can provide the complete maintenance ration for the mares from approximately May onwards and as such is a highly important factor in the management of any stud.

Horses by their very nature tend to poach land to a far greater extent than other classes of stock. For this reason, the lightest and/or best drained land available should be choosen for stud paddocks. Light land on a chalk sub-soil, is ideal from this point of view, so long as the land is not stony, which would cause some shallow footed horses to be constantly lame. However, heavy land is generally more fertile, but has the disadvantage of becoming badly poached during the winter with very muddy gateways.

Paddocks should be level, particularly where foals and youngstock in general are concerned, as accidents can happen, (especially when the ground is slippery), if horses moving fast downhill are unable to pull up when they meet the fence at the bottom.

On most Thoroughbred studs, the visiting mares start to arrive from the beginning of February onwards. At this time of the year and through to early April, there are usually some periods of very cold winds, which will tend to prevent mares from cycling normally. For this reason, thought should be given to shelter when laying out paddocks. Where possible, old belts of trees and high hedges should be left intact; but where none exist, quick-growing, non-poisonous varieties should be planted, at sufficient distance outside the paddock fence so that the horses cannot reach them. Copper Beech makes an ideal hedge as it gives shelter during both summer and winter, looks beautiful and is non-poisonous – but is rather expensive. Ideally, shelter should be established on all four sides of the paddock but failing this at least against North and East winds. Close boarding at particularly vulnerable points can be used until shelter belts are established.

As already discussed, horses tend to poach land badly, therefore land used for paddocks should be well-drained. As far as possible, it is usually advisable to tile drain any ditches and fill them in altogether. This will prevent accidents happening, particularly where young foals are concerned. Failing this, ditches and ponds should be very well fenced with at least three, and preferably four or more, rails, as young foals will often climb through or roll under a two-rail fence.

To save labour, water should be laid on to all the paddocks. The troughs should have automatic filling devices and have some means of easy drainage in the bottom to facilitate regular cleaning. Water troughs are better sited away from gateways, otherwise they will only add to the poached condition of this part of the paddock in wet weather. As far as possible, they should not be placed under trees, otherwise you will be constantly scooping leaves out, especially during the autumn, but at the same time they should be sited where there is some shelter.

When selecting grass mixtures for pastures which are to be used mainly as stud paddocks, the following points must be taken into consideration:

(1) An early bite is essential, particularly on Thoroughbred studs, where the mares start coming in from the end of January and most have gone home again by mid-June.

(2) Horses are very selective grazers; therefore, palatability is of great importance and strains of grass should be selected for paddock mixtures with this in mind. Nearly all grasses are palatable when young, but some become less palatable with age. Leafy varieties have a higher protein content, with lower fibre and are therefore more digestible. Cocksfoot, except when it is very young, is too coarse and tends to be rejected by horses.

Horses have the habit of cropping the grass almost to ground level. Care should therefore be taken not to include too much clover in any paddock mixture, as under heavy grazing conditions, this will tend to grow away at the expense of the grass – providing a low overall yield of herbage per acre.

A grass mixture containing several different varieties, including herbs and clover, is ideal for horses but the accent must be on early varieties for public studs. For horses, quality not quantity is the key factor.

(3) Paddock grasses must be able to withstand poaching to a large degree; therefore, they must have good tillering ability – that is, the ability to give off new shoots at their base and thus spread out at ground level.

(4) Grasses which also exhibit speed of recovery after grazing are desirable. The type of land will also influence the paddock mixture. When in doubt, your local A.D.A.S. officer should be consulted.

The fencing on public studs should be post and rails or high stock proof hedges. The rails round the stallion's paddock can be painted white for effect, but all the other rails are better creosoted to prevent the horses from eating them. With the current high prices of this type of fencing, constant applications of creosote is well worthwhile.

As already discussed, wherever possible three rails should

be put up to prevent young foals from getting out. Double fencing will help to prevent the spread of disease and will also help to prevent horses from leaning on the fence in an effort to reach their neighbours on the other side. Walks can be constructed between the paddocks, so that the teaser can be taken down to try the mares, with easy access to every paddock. Trying bars can then be constructed in the paddock fences. Stallion paddocks can be close boarded (see plate 15) but where a paddock can be found which does not look out on to the mares, the stallion paddock can be fenced with post and rails to a height of at least 6 ft. for safety. All paddock corners should be rounded to prevent accidents.

Wooden gates are probably safer than metal gates, inasmuch as they will tend to break if a horse puts a leg through them. Where metal gates are used, they should have their bars running vertically rather than horizontally, to help to prevent accidents. Wooden gates last longer if they have metal bands on to prevent the horses chewing through them. Ease of opening and closing is essential on a stud, and a wicket gate type fastening is ideal. In order to ensure that the horses do not open the gates themselves, a chain should also be fitted, together with a staple and metal pin.

Gates should be located, ideally where there is direct access on to a hard road. The area should be extremely well-drained to cut poaching down to a minimum. Corners of paddocks should be avoided as far as possible, so horses do not get kicked when they are being caught up and are congregating round the gate.

In order to get good grass, the soil must be in good condition, as well as being well-drained. This means there must not be any deficiencies of lime, phosphates, potash or nitrogen. Excess lime can lead to deficiences of other minerals – e.g. phosphorous, with resulting bone troubles. The application of Basic Slag (phosphorous) and lime is now thought to produce a high Molybdenum content in the grass, leading to a copper deficiency in the grazing animals and a possible softening of their bones. Lime, phosphates, potash and nitrogen are all washed out of the top-soil over a period of time. Where there is any difference in vegetation over the area of your paddocks, e.g. moss, buttercups which might lead you to believe that there are uneven patches of acidity, it may be advisable to buy a soil indicator outfit and test the paddocks yourself. These can be obtained from your local Agricultural Development and Advisory Service office and are quite easy to use.

As with lime and potash, phosphates tend to be washed out of the top soil, therefore regular testing of paddocks should be carried out, ideally once every three to five years. Normally many of the nutrients are replaced by the grazing animals but where droppings are removed regularly, deficiencies may occur. A phosphate deficiency can cause soft bones, which may result in splints, etc. Phosphate is essential for root development and benefits clovers. Which in turn help the nitrogen level. Potash is necessary for good grass production.

Nitrogen is leached out of the soil at a far greater rate than any of the elements already mentioned, and needs renewing annually. High applications of nitrogen will lead to heavy crops of grass, quite unsuitable to the needs of horses, but where an early bite is required, a light top dressing of a nitrogen fertilizer – i.e. not exceeding 45–50 units of nitrogen – will provide the necessary level of growth, if applied in the early spring just as the grass is beginning to grow.

Ideally, a liberal application each autumn of ordinary farm-yard manure – not poultry manure, which is too high in nitrogen – together with lime when necessary, will supply all the nutrients necessary to establish and maintain stud paddocks in optimum condition. In the spring and between grazings the paddocks should be harrowed, ideally using ripper type harrows, to pull out any dead grass and spread the remaining manure. After harrowing, the paddocks should be rolled with heavy grass rollers – the Cambridge variety are really too light for grassland.

Throughout the season, the droppings should be removed. The paddocks should be trimmed when necessary, where they are not being inter-grazed with cattle.

In order to prevent over use, the paddocks are better grazed for alternate periods of 2–4 weeks and one or two should be set aside in annual rotation for hay. Basically, nature cannot stand high concentration of any single species in a given area and under these conditions disease usually occurs. This will usually manifest itself in the form of lice, worms and general unthriftiness of the stock. Overstocking applies equally to buildings as well as land.

It is imperative therefore that the total concentration of stock is kept down; mixed stocking is practised, e.g. horses and cattle; land and buildings are rested for periods of months rather than weeks, between periods of heavy stocking.

With these factors in mind, one cannot go too far wrong!

16

THE PROBLEM OF WORMS

Worms are parasites. A parasite is an organism which lives on or in another organism (in this case the horse) deriving its nourishment from the host animal, but without actually killing it. If the parasite does destroy its host it is defeating its own object – i.e. it is depriving itself of a place to live and reproduce its species. Therefore where an animal dies as a direct result of a worm infestation it is a grave error on the part of the worms and not the inevitable result in a natural course of events.

Why then should we worry too much about worm parasites? The reason is that horses which have not previously come in contact with worm infections are not immune to their effects – while other immune but infected horses are capable of infecting all other horses with which they are turned out. The external signs of parasitism are:

(i) Loss of weight
(ii) Loss of condition, poor coat, etc.
(iii) Debility
(iv) Anaemia
(v) Colic
(vi) Death

At all costs one should avoid sending mares home suffering from any of the effects of parasitism; this includes lice as well as worms. Affected mares will give a stud a bad name in a very short space of time, which will take several years to redeem.

The worm parasites usually involved are:

(i) Red worms – large and small
(ii) Round worms
(iii) Strongyloides
(iv) Tapeworms
(v) Seat worms
(vi) Lungworms
(vii) Liver Fluke
(viii) Bots

A complete list of these worms with their various degrees of pathogenicity can be seen on page 141.

The strongyles are by far the most common and important species and this is the family of the Red Worm. The adults live in the intestines, where the females lay their eggs, which pass out in the droppings on to the grass; they hatch out into larvae and climb up the blades of grass, where they hope to be eaten by their particular host animal. Any larvae which are fortunate enough to be eaten, pass, with the grass, into the animal's stomach. If the animal is their particular host species, they will survive, but if not they will be destroyed. Larvae which survive will migrate through the intestinal walls, travelling up the mesenteric arteries while the small strongyle remains in the walls of the intestine. The mature larvae then eventually return to the intestines as adults where the females commence egg laying. (See Fig. 27.) The migrating larvae can cause aneurisms to form at the root of the mesenteric arteries, which can result in sudden death, or damage to part of the gut resulting in chronic intermittent colic.

Another type of worm most commonly associated with young horses, is the ascarid (Round Worm). These can easily be seen in the droppings of infected stock; they measure up to 14 inches in length. In the case of very large infections, these worms may cause intestinal obstructions and sometimes

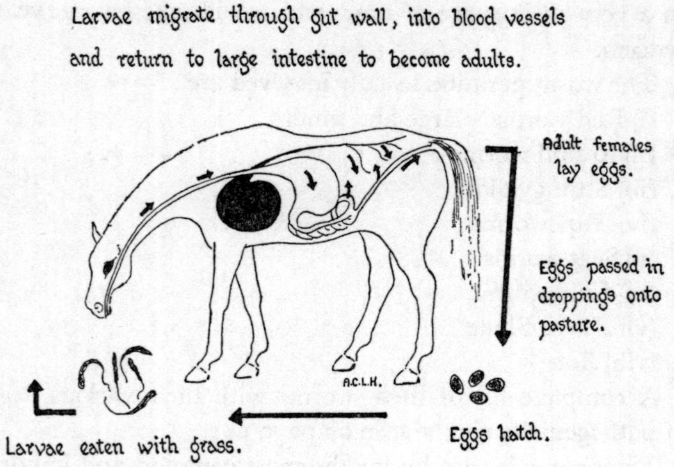

Fig. 27. The life cycle of the red worm

also perforation of the bowel, which results in a fatal peritonitis.

Strongyloides are common in young foals; they appear to be self-limiting as they are seldom seen in the adult horse; they are therefore of little consequence, although some authorities maintain that they can be the cause of early scours.

Tapeworms seldom cause much trouble in horses unless there is a large infestation.

The eggs of the Seat Worm (Threadworm) are seldom, if ever, found in faece samples sent for analysis, due to the fact that the female worm lays its eggs around the anus and not in the intestine, as do most of the other equine worm parasites. The eggs usually set up an irritation causing the horse to rub its tail and hindquarters.

Lungworms are often found in the donkey, which appears to be their natural host, they seldom produce symptoms unless accompanied by a secondary infection. In the horse, however, there are sometimes symptoms of coughing and broken wind – even in the case of small infections. Where lungworm occurs in the horse it usually originates from an infected donkey. (See Fig. 28.)

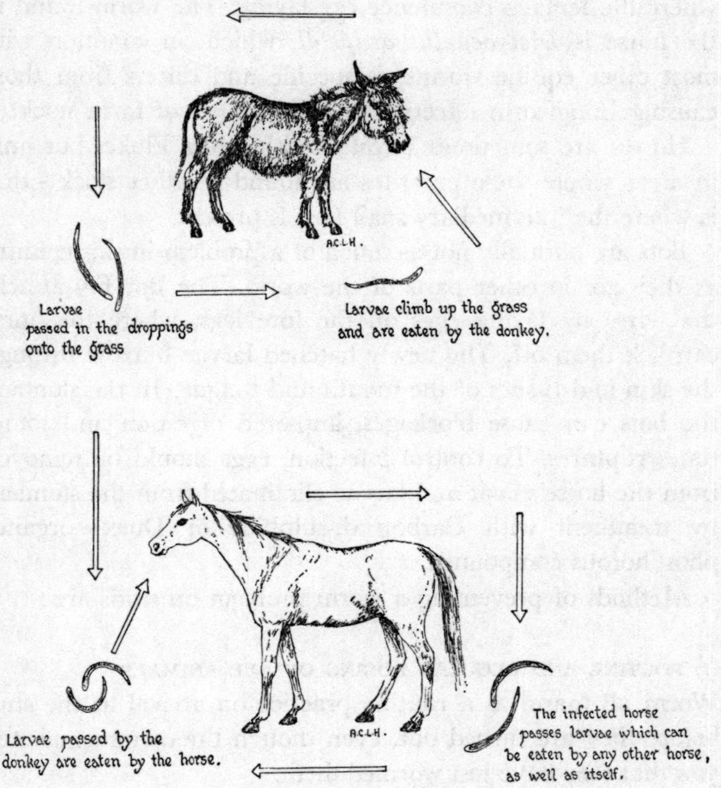

Fig. 28. The life cycle of the lungworm

The life cycle of the Lungworm differs from those already mentioned in that, although the female lays her eggs in the intestines, most of these hatch before they are passed in the faeces. On the pasture they mature and, when swallowed by a horse or donkey, pass down into the intestines; from the intestines they enter the lymphatic system and thence reach the right side of the heart. From the right venticle they become blood-borne and migrate to the lungs; here they break out into the air passages and eventually reach the windpipe when they are either coughed up and swallowed or pass up in the mucous and are swallowed, returning to the intestines, where the females commence egg laying. The worm found in the horse is *Dictyocaulus arnfieldi*, which, in common with most other equine worms, is specific and differs from those causing Lungworm infection in other classes of farm stock.

Horses are sometimes infected with Liver Fluke, but only in areas where these parasites are found in other stock – that is where the intermediary snail host is present.

Bots are normally not as much of a problem in this country as they are in other parts of the world. The Bot Fly attacks the horse by laying eggs on the fore legs, where the horse can lick them off. The newly hatched larvae burrow through the skin and tissues of the mouth and tongue. In the stomach the bots can cause blockages, impaired digestion and sometimes ruptures. To control infection, eggs should be removed from the horse's coat and larvae eliminated from the stomach by treatment with Carbon di-sulphide or Duex – organo-phosphorous compounds.

Methods of preventing a worm problem on studs are:

(I) ROUTINE AND REGULAR DOSING OF ALL ANIMALS

Worm all mares as a routine practice on arrival at the stud before they are turned out, even though the owners may tell you that they have just wormed them.

Get *all* animals up from the paddocks every six to eight weeks for routine worming. At the same time, worm all other animals on the stud including the stallions and any foals over eight weeks old. In this way, all the horses will be kept free from egg-laying adult worms in the intestines, and your pastures will remain relatively clean.

It is only necessary to dose adult stock against strongyles unless you have reason to believe that an individual is also infected by another species of worm. All youngstock must, however, be dosed against both ascarids (Roundworms) and strongyles.

(II) PADDOCK MANAGEMENT

Ideally, all droppings should be removed from the paddocks daily, but with the present high cost of labour this is not always a practical proposition. However, every effort should be made to remove the droppings from the smaller or most heavily used paddocks, otherwise they will tend to become 'horse-sick'.

From the point of view of worm control, paddocks ideally should be rested for the six-month winter period, when frost will drastically reduce the numbers of surviving larvae.

Horses are very selective in their grazing habits and will only graze areas which have not been contaminated by their own droppings. Therefore large areas of a paddock will tend to be spurned by the horses with the result that unless the grass is kept topped regularly it may grow very long and coarse in places. An alternative method to this system is to graze a herd of de-horned or poll cattle with the horses. They are less selective in their grazing habits and will destroy most of the worm eggs, as well as keep down the long grass. Ideally, one should maintain a grazing rate of five horses to twenty-five cattle, but this is not usually possible, except for very large acreages (plate 13).

The practice of harrowing paddocks, to scatter the droppings, only has the effect of contaminating larger areas and reducing the clean area available to the horses. Ideally, and if at all possible, droppings should be removed from the paddocks at least once a week. When paddocks get in a very bad state, deep ploughing and re-seeding is often the only possible remedy.

(III) ROUTINE ANALYSIS OF DROPPING SAMPLES

All horses should be wormed or have their droppings tested *before* they are turned out. Where possible, the worm count can be carried out on the stud, to eliminate the usual delay between collection of the sample and testing. It is preferable to test all animals on the stud every other month as some varieties of worm become immune to a particular drug and infected horses will continue to pass worm eggs, even though they are being treated regularly. The Lungworm and Liver Fluke counts are probably beyond the scope of an average stud and samples for these counts should be given to your veterinary surgeon, or sent to a laboratory for testing.

Equipment needed for the worm egg count test

(a) A microscope with a low-powered lens, i.e. (x 8 eyepiece and x 4 objective lens), is ideal; these can be bought very cheaply from most chemists and camera shops and come with full operational instructions.

(b) A McMaster slide which can be obtained from your veterinary surgeon or direct from: Hawksley and Sons Ltd., Peter Road, Lancing, Sussex. (See Fig. 29.)

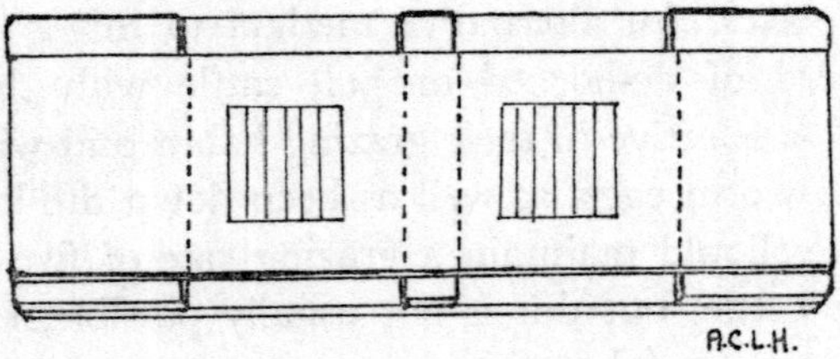

Fig. 29. A McMaster slide

(c) A fine mesh nylon tea strainer obtainable from most hardware shops.

(d) A graduated 20 ml. syringe – obtainable from your veterinary surgeon.

(e) A Pasteur pipette (or any other similar type) fitted with a rubber teat, obtainable from most chemists.

(f) A supply of cooking salt.

(g) A few small honey jars (or similar containers).

Method of carrying out the test

Use Fig. 30 as a guide:

2 grams of faeces are approximately the size of a small walnut – accuracy in weighing is unnecessary, as all we are

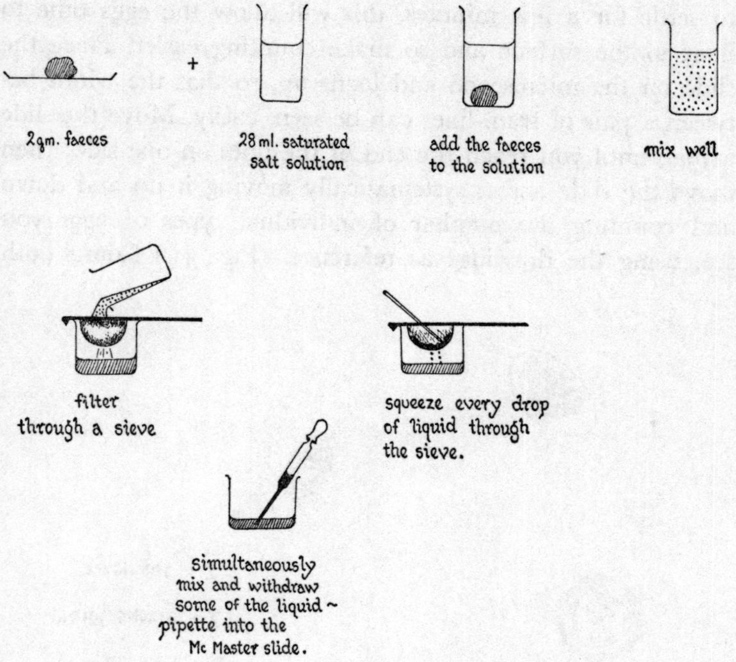

Fig. 30. The worm egg count test

interested in is whether the individual horse has worms; therefore err on the heavy side rather than underestimate the size of the sample.

Make up a saturated salt solution in advance. To do this, simply add common salt to water until no more will dissolve out and some un-dissolved salt remains in the bottom of the container. Let the salt particles settle out before use.

Withdraw 28 ml. of the saturated salt solution, using your graduated 20 ml. syringe. Place the solution into an empty container and add the faeces sample. Mix well. Using the fine nylon sieve, filter the sample, into a clean jar, mashing it well in the sieve, to squeeze every drop of liquid out.

Then, mix and withdraw the liquid; fill one chamber of the slide, repeat and fill the other chamber. Leave the slide to settle for a few minutes, this will allow the eggs time to float to the surface and so make counting easier. Place the slide on the microscope and focus up, so that the width between a pair of tram-lines can be seen, easily. Move the slide across, until you reach the end of the lines on one side; then move the slide across systematically moving it up and down and counting the number of individual types of eggs you see, using the drawings as reference. (Fig. 31.) Count both

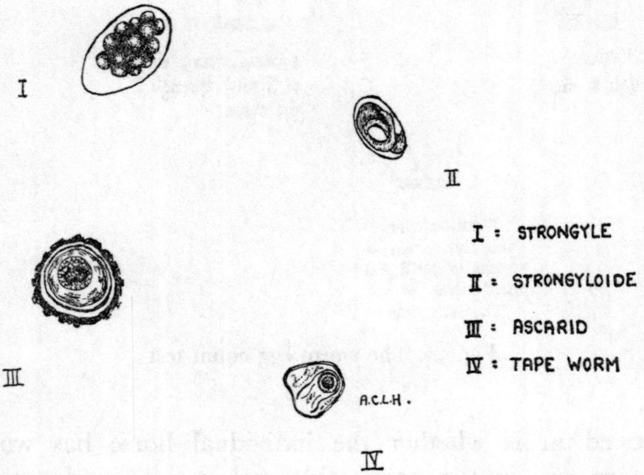

Fig. 31. Individual worm eggs

chambers. The total number of eggs of each variety found x 50, will give you the number of eggs per gram of faeces.

Any animal with a worm egg count should be treated. The actual count will vary with the numbers of egg laying adult worms present in the intestine at any one time and to a large extent on the time of year: egg laying takes place more actively during the spring and early summer.

Consult your veterinary surgeon as to the most suitable drug to use.

Common Internal Parasites of the Horse

Group	Common Name	Proper Name	Degree of Pathogenicity
1 (a)	Large red worms	*Strongylus vulgaris*	++++
		Strongylus equinus	++
		Strongylus edentatus	++
(b)	Small red worms	*Trichonema*	+
		Triodontophorus	+
		Trichostrongylus axei (common also to cattle and sheep)	+
(c)	Round worm	*Parascaris equorum*	+++
(d)	Strongyloides	*Strongyloides westeri*	+
(e)	Seat worm	*Oxyuris equi*	±
2	Tapeworm	*Anoplocephala pertoliata*	±
3	Lungworm	*Dictyocaulus arnfieldi*	++
4	Live fluke	*Fasciola hepatica*	±
5	Bots	*Gastrophilus*	±

17

INSURANCE

If you own a large number of mares, it often pays not to insure them, as you can afford to loose one occasionally, the loss usually not amounting to more than the total annual premium. It is, however, a great mistake to insure one or two mares out of a large number and leave the others uninsured as you can be certain that if anything is going to die it will be one of the uninsured mares. If you only own a few mares it always pays to insure every one.

This does not apply to the stallion, as he is, in most cases, the primary reason for the existence of the stud, and he would have to be replaced quickly, probably at considerable expense, should he die or become incapable of getting his mares in foal.

Basically there are four types of stallion insurance:

(i) All risks of mortality;

(ii) Congenital infertility;

(iii) Permanent infertility;

(iv) Loss of income.

(i) ALL RISKS OF MORTALITY

As stated, this insurance covers the horse against all risks of death, but not against accidental injury which does not actually result in death. Nor does it cover infertility.

(ii) CONGENITAL INFERTILITY

This type of insurance only relates to the *first season* horse and covers the owner against his stallion proving either infertile, partially infertile or incapable of serving mares, during his first season.

A hundred per cent of the insured sum is returnable in the event of the first season sire proving to be a hundred per cent infertile.

Infertility in this case means a lack of libido; uninduced refusal to serve mares; or any abnormality of the semen which results in markedly reduced ability to fertilise mares, under normal conditions of stud management.

The fertility percentage of a stallion is calculated within sixty-five days from the end of the covering season and is based on veterinary certificates of pregnancy, obtained from the visiting mare owners.

Seventy-five per cent of the insured sum is returnable in the event of the stallion proving only ten per cent fertile.

Fifty per cent of the insured sum is returnable in the event of the stallion finishing the season with a fertility percentage between thirty-three and a third per cent and ten per cent.

In order to make a claim under this type of cover, it is usually stipulated that the stallion has a minimum of twenty mares booked to him, unless a smaller number has been agreed beforehand. The visiting mares must be clean and in sound breeding condition.

This insurance does not usually cover against infertility resulting from accidental external injury to the stallion, nor does it cover against infertility arising from genital infection in any of the visiting mares.

(iii) PERMANENT INFERTILITY

This covers the loss which may occur in the event of the insured stallion becoming permanently impotent, infertile or incapable of serving mares, as the result of an accident, illness or disease occurring during the period of the policy.

This insurance does not give cover for loss arising from the death of the stallion.

(iv) LOSS OF INCOME

This policy covers the stallion owner or shareholder in a stallion against the actual loss of income which would result, in the event of the stallion's failure to serve or complete service of his nominated mares, in any particular covering season; this includes physical causes and also death.

'Complete service' in this case means that the horse must be available to the nominated mare for a period of seventy-five days from the time of first service or to July 15th (end of the covering season), whichever period is the shorter.

The insured stallion must be unfit for stud duties for a period of at least ten days to constitute a claim.

It is usually agreed that the insured person should try to obtain the services of a substitute stallion and that the underwriters should only pay the difference, between the fees obtained from the substitute stallion, and those which would have been obtained from the original stallion, had he been able to complete a normal stud season. There is usually a return of a part of the premium in the event of no claim.

A stallion owner may insure himself only against loss arising from an outbreak of Rhinopneumonitis. In this case, cover is given against loss of income resulting from an outbreak on the owner's stud, when mares will be unable to come in to visit the horse, or where an outbreak has occurred on another stud and a visiting mare from that stud is unable to come in for service, due to the risk of infecting the stallion and mares on the insured owner's stud. This can also be extended to cover other diseases, such as strangles, if necessary. A condition of this insurance is usually that there has not been a case of Rhinopneumonitis on the premises during the previous six months.

In the case of a private stud

If a case of Rhinopneumonitis is diagnosed at a mare owner's stud, it means that *all* the mares on the stud will have to be isolated for a period of at least two months and if this happens during the covering season, it will mean that the owner will not be able to take up his nominations for that year, and unless he is able to sell them, will be under an obligation to pay at the end of the covering season.

If, as a direct result of an outbreak of Rhinopneumonitis on the mare owner's stud, a mare cannot visit the stallion until after June 1st, in any one year, it is usual for fifty per cent of the sum insured to be paid.

In the case of a public stud

Where the mare is away at a public stud and there is an outbreak of Rhinopneumonitis on that stud, there will be a ban on the movement of *all* mares both into and out of the stud. The mare owner may find himself involved in the payment of keep charges well in excess of normal expectations, through no fault of his own.

This policy will either compensate the mare owner direct or, probably a better idea, it will compensate the stallion owner if he insures his stud against an outbreak of Rhinopneumonitis. In this case he does not charge the mare owners any extra keep but simply claims from the insurers.

A condition of insurance is that apart from no outbreaks having occurred on the stud during the previous six months, no mares must have been vaccinated against virus abortion during the previous six months, which might be relevant where American mares are concerned.

STUD OWNERS LEGAL LIABILITY INSURANCE

This covers the insured person against any claims that he might become legally liable to pay in respect of horses which are the property of others and are in his temporary care, custody and control or that of his employees.

Care should be taken that the amount insured will in fact cover the value of the most expensive mare visiting the insured's stallion. This policy covers the death or permanent injury of any horse that is the property of others, resulting from a negligent act, error or omission on the part of the insured or his employees. It excludes death due to infectious or contagious disease. It also excludes any animals partly owned by the assured or his employees. It does, however, give cover against legal costs, providing the insured person does not admit liability in the first place.

When this insurance is taken out by a stud, the fact must on no account be made known or publicised, for obvious reasons.

18

STUD RECORDS

It is absolutely essential that accurate records should be kept of matters relating to all mares. The usual records to be found on most public studs are:

(i) A day book;
(ii) Stud groom's record book;
(iii) Trying book or index cards;
(iv) Foal record sheet;
(v) Wall chart – to record trying and covering;
(vi) Chart recording days from last service;
(vii) Performance records of progeny;
(viii) Paddock records;
(ix) Ready reckoner;
(x) Account records.

(i) THE DAY BOOK

This is just a diary of the daily happenings on the stud. A hard-backed exercise book is quite satisfactory; the book should be left in the stud office where it can be used by the stud groom, second man, secretary or manager. The type of information it should contain is the arrival and departure of all stock on and off the stud, with some note of the means of transport i.e. owner's box, stud horsebox, etc., which is essential for making up accounts later. All foalings, coverings, blacksmith's visits, with a note of the mares attended, and likewise veterinary visits for worming, vaccination markings, etc., and any other relevant details which should be entered as they occur.

(ii) STUD GROOM'S RECORD BOOK

These can be obtained from Tindall & Sons in Newmarket. The hard-backed, bound volume is probably the most satisfactory for the stud groom, as it will stand up to harder wear than the loose-leaf variety.

Most managers will find it useful to keep a small loose-leaf copy of the stud groom's book and carry it round all the time. It is a great advantage if you record the coverings, tryings and brief details of pedigree and progeny performances on the reverse of the individual pages. You then have something of a pocket bible, which gives all the relevant information of every mare at a glance and is invaluable when speaking to owners on the telephone at home, or when showing people round the stud.

(iii) TRYING BOOK OR INDEX CARDS

The trying book which is kept by the stud groom or manager can be obtained from Tindall & Sons mentioned above or an index card can be kept on each mare see Fig. 32. The following annotations can be used:

AR : arrived

D : due to foal

F : foaled

X : in-season

Ⓧ : covered

P : injected with prostaglandin.

 : mare tested in-foal, the colour of the line can be changed when the mare returns home.

The stud groom should also keep an ordinary exercise book, in which he can note the mares he tries each day and record their behaviour at the trying board.

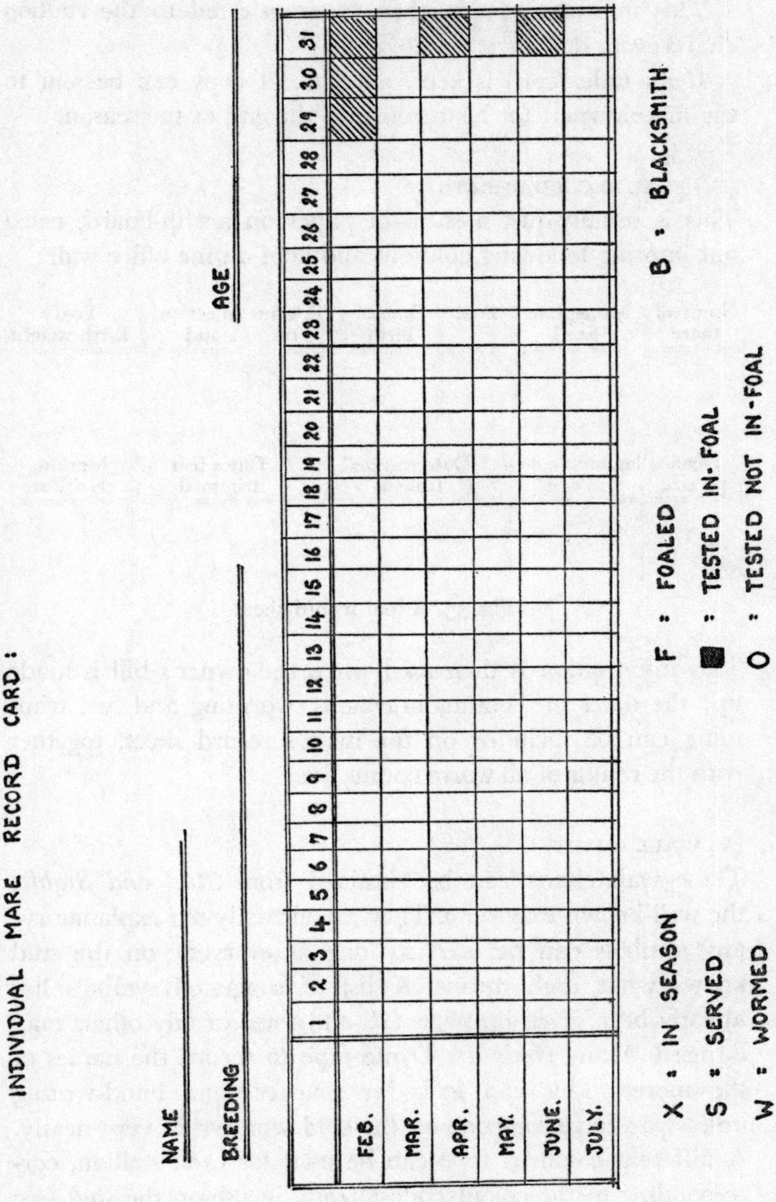

Fig. 32. Individual mare record card

This information can then be transferred to the stallion charts every day.

If an index-card is kept, a photostat copy can be sent to the mare owner, for his records, at the end of the season.

(IV) FOAL RECORD SHEETS

This is usually just a sheet of paper on a clip-board, ruled out into the following columns and kept on the office wall:

Name of mare	Nameplate fixed	Foal	Date of Birth	Tetanus Injection		Foal's Birth weight
				1st	2nd	

Headcollar supplied		Date wormed (tubed)	Dates feet trimmed	Marking certificate
1st size	2nd size			

Fig. 33. A foal record sheet

This information is then used when the owner's bill is made up; the dates of Tetanus injections, worming and feet trimming can be included on the mare's record sheet, together with the results of all worm counts done.

(V) WALL CHART

These wall charts can be obtained from *Stud and Stable,* the well-known magazine. They are virtually self explanatory; any symbols can be used so long as everyone on the stud knows what each means. A list of suggested symbols has already been given on page 148 and these or any others may be used. Many studs use Dymo tape to record the names of the mares. This can look far smarter than hand-writing unless you have someone on the stud who writes very neatly. A different coloured tape can be used for each stallion, corresponding to the colour code already in use on the stud (see page 33). Any mistakes on the chart can be very easily erased by painting them out with 'Snopake', obtainable from most stationers. Some studs like to keep the wall charts concealed, as some of the information they contain could possibly be considered confidential.

(vi) CHART RECORDING DAYS FROM LAST SERVICE

See Fig. 34. This is a very useful chart to keep near the telephone, as it gives an 'at-a-glance' guide to the number of days any particular mare has gone from last service. When any mares on the list 'break' (come in-season again), they can be crossed off the list. The list itself should be renewed once every ten days or so. When there is more than one stallion on the stud, a coloured dot (in accordance with the colour code used for each horse), can be put against each mare's name as a guide.

NAME OF MARE	1st	2nd	3rd	4th	5th	6th	7th	8th	9th	10th	11th	12th	13th	14th	15th
Goldie	39	40	41	(42)	43	44	45	46	47	48	49	50	51	52	53
Bracken	32	33	34	35	36	37	38	39	40	41	(42)	43	44	45	46
Sally	25	26	27	28	29	30	31	32	33	34	35	36	37	38	39
Jet / Misty	~~22~~	~~23~~	~~24~~	~~25~~	~~26~~	~~27~~	~~28~~	~~29~~	~~30~~	~~31~~	~~32~~	~~33~~	~~34~~	~~35~~	~~36~~
Star	(21)	22	23	24	25	26	27	28	29	30	31	32	33	34	35
Tinkerbell / ~~Arrow~~	~~16~~	~~17~~	~~18~~	~~19~~	~~20~~	(21)	~~22~~	~~23~~	~~24~~	~~25~~	~~26~~	~~27~~	~~28~~	~~29~~	~~30~~
~~Penny~~	14	15	16	17	18	19	20	(21)	22	23	24	25	26	27	28
Marcia	13	14	15	16	17	18	19	20	(21)	22	23	24	25	26	27
Topsy	10	11	12	13	14	15	16	17	18	19	20	(21)	22	23	24

Fig. 34. Last service chart

(vii) PERFORMANCE RECORDS OF PROGENY

Unfortunately, one cannot buy this chart already ruled out, but it does not take very long to make one yourself. All you need is some white card.

Rule the card out into the following headings:

Age	Name	Sex	Dam	Sire of Dam	Trainer	Breeder

Fig. 35. Performance chart

Leave approximately twenty blank spaces for the race or show results, etc.

This chart should be kept in a position where it can be seen by *all* the staff, as it can create great interest and enthusiasm especially if the horses are doing well.

(VIII) PADDOCK RECORDS

These paddock charts should also be kept where it is easily accessible to everyone on the stud. You will need a map of the stud and as many plastic envelopes as there are paddocks. Index cards are then kept in the envelopes on which the individual paddock management can be recorded, e.g. fertilizer treatment, soil analysis, periods grazed and rested, etc. These envelopes can also be used to record the mares in the paddocks. The names of the mares are printed out on Dymo tape, which will easily stick to the envelopes and just as easily peel off and re-stick on to another envelope when the mares are moved to a different paddock. This, then, becomes an invaluable guide to the whereabouts of every single mare on the stud. Different coloured tapes or threads should join the envelopes to the relative paddocks for easy reference.

A map of the layout of the stud loose-boxes should be kept. Covered with perspex, which can be written on with a grease pencil, or Dymo tape, it will give a guide to the position of every stabled mare on the stud. (See Plate 32.)

(IX) READY RECKONER

This is almost essential when you are working out the owners' bills. Place the various standard charges in columns across the top of the page. Down the page, list the number of months in one section and then in a separate section, the number of days, i.e. one to seven. Work it all out, preferably using an adding machine.

It is suggested that keep charges should be sent out once a month, so that owners can query their accounts before the end of the season and have a chance of paying before the bill gets too huge.

In the case of pure-bred stock, it sometimes helps to galvanize people into paying their bills, by retaining the covering certificate until you have received their money.

(x) ACCOUNT RECORDS

Fig. 36 shows the best method of presenting the account.

Tel. No. STUD ADDRESS ...
....
....

....197...
....
.... V.A.T. Reg. No. ......

Season .... Nominations to		
Keep of .... from .... to		
.... weeks .... days @ .... per week/day		
Keep of .... from .... to		
.... weeks .... days @ .... per week/day		
Groom's Fee		
Blacksmith		
Worm Control		
Telegrams / Telephone		
Sitting Up .... Nights @ .... per night		
Transport		
Miscellaneous		
.... covered by		
Last service date		
.... covered by		
Last service date		
Total £		
Plus V.A.T. @		
Total Due £		

Cheques should be made payable to ....

USEFUL ADDRESSES – referred to in the text:

Messrs. Weatherbys,
Sanders Road,
Wellingborough,
Northamptonshire.

Hunters Improvement and National Light Horse Breeding Society,
8 Market Square,
Westerham,
Kent.

Dr M. C. Round,
The Garrod Laboratory – Parasitological Service,
121 Girton Road,
Cambridge CB3 OLS.

Tindall and Son, (Specialists in stud stationary and printing)
50 High Street,
Newmarket,
Suffolk.

R. P. Faulkes Esq. (Specialist horse photographer)
49 Park Lane,
Newmarket,
Suffolk.

Stud and Stable,
149 Fleet Street,
London EC4A 2BU.

National Foaling Bank,
Meretown Stud,
Newport,
Shropshire (Newport 811234).

INDEX